Automotive Systems Engineering

Markus Maurer · Hermann Winner
Editors

Automotive Systems Engineering

Springer

Editors
Markus Maurer
Institut für Regelungstechnik
Technische Universität Braunschweig
Braunschweig
Germany

Hermann Winner
Fachgebiet Fahrzeugtechnik
Technische Universität Darmstadt
Darmstadt
Germany

ISBN 978-3-642-36454-9 ISBN 978-3-642-36455-6 (eBook)
DOI 10.1007/978-3-642-36455-6
Springer Heidelberg New York Dordrecht London

Library of Congress Control Number: 2013935997

Printed on acid-free paper

Springer is part of Springer Science+Business Media (www.springer.com)

Preface

Systems Engineering has long been established in the fields of avionics and software development. In the automotive sector too, we are witnessing an increasing number of systems engineering approaches. Nonetheless, a distinct profile has still to be developed that clearly defines its limits and interfaces with other development and research areas and gives a specific identity to automotive systems engineering. Consequently, conferences and publications offering opportunities to exchange experience on this topic are also lacking. Despite the fact that many scientists and practitioners are working on Automotive Systems Engineering issues, a pinpointed community has not yet emerged that covers at least the major areas involved. Automotive Systems Engineering pools many disciplines, from control engineering, test development, development of mechanical engineering, electrical and electronic hardware and software. And through ergonomics, issues of human physiology and psychology also permeate through the driver-vehicle system.

Automotive Systems Engineering combines a multiple of disciplines, but being "sandwiched" between so many different stools its position is also weakened. Personal talks between the editors on this situation lead to a workshop where ASE-related activities were presented and an intensive exchange of experience was initiated. The successful outcome resulted in this publication with the ambition of encouraging an exchange of experience beyond our two institutes. The topics covered in each chapter have their roots in doctoral theses still ongoing or nearing completion. They present the state of science centering on development methodology and with a limited application focus. A clear heterogeneity is visible both within and between the institutes, although it would be feasible and also desirable to reach more harmonised approaches. The perspective pursued with this publication, therefore, is to enhance standardisation, both in the terminology used and in prevalent definitions for methods—a common process when building communities. By means of workshops and specialised publications, we intend to advance the community-building process together with a wider range of partners from academia and industry and trust that the present work will provide an appropriate foundation.

Alongside chapters presenting scientific studies, the editors' subjective perspectives on the topic are added, leaning on talks and lectures and their background experience in academia and industry.

We wish our readers stimulating reading and look forward to receiving a wide spectrum of feedback to help us tread the path towards an automotive systems engineering community.

Markus Maurer
Hermann Winner

Contents

Part III Functional Safety

Part IV Evaluation of Perception Capabilities

Part V Functional Testing

Part I
Background and Definitions

Chapter 1
Challenges of Automotive Systems Engineering for Industry and Academia

Hermann Winner

1.1 Preface

After many years working on systems development in industry and academia, I have developed a personal view of the subject which is reflected in the present paper. This is not based on systematic research, has no claim to be representative and may be highly subjective. It is the product of my experience in a number of predevelopment projects and an initial development for volume production at a tier-one supplier, as well as university research into functional and methodological aspects of Systems Engineering.

1.2 Definition of Terms

1.2.1 System

A group of elements which are relevant (and not merely useful) for achieving a purpose, which interact with each other, and which have a structure within predefined boundaries.[1]

While this and similar definitions allow us to characterise a system, in practice it is particularly difficult to define the boundaries of the system. In a transport system, a road vehicle (possibly together with a driver) is an element of the system.

[1] As there are many different definitions for the term "system", major parts of the present definition will unavoidably overlap with one or several of them. No claim to originality is made, therefore, in regard to this or any subsequent definitions used.

H. Winner (✉)
Fachgebiet Fahrzeugtechnik, Technische Universität Darmstadt,
Petersenstrasse 30, 64287 Darmstadt, Germany
e-mail: winner@fzd.tu-darmstadt.de

M. Maurer and H. Winner (eds.), *Automotive Systems Engineering*,
DOI: 10.1007/978-3-642-36455-6_1, © Springer-Verlag Berlin Heidelberg 2013

For an automotive manufacturer the car is the system, for the unit responsible for a module—whether a vehicle manufacturer or a module supplier (e.g. for the brakes)—the module is the system, where elements of this system can in turn be elements of another system, e.g. the data communication system or a system which functionally serves a different purpose, such as an Adaptive Cruise Control. A well-chosen system boundary contains all the elements which are relevant to the purpose and amenable to influence, with a minimum level of complexity.

1.2.2 System Design

Concept, specification, implementation, verification and validation of a technical system for achieving a specific (and mostly functional) objective. Project management can be an integral or complementary element.

This definition consciously follows the familiar V-model, which describes development as a phase of project definition descending to the point of implementation and subsequently rising again through a test phase to the highest level. How far project management is part of system design is a matter of opinion. This will, however, be the case where the system design process involves both functional and organisational roles, e.g. if there is strict separation of responsibility for implementation and responsibility for approval.

1.2.3 Systems Engineering

Goal-independent methodology for developing mostly complex systems.

Unlike system design, which serves to achieve a purpose, Systems Engineering is a methodological approach which assists in achieving high development quality, in terms of meeting quality, cost and timetable goals. However obviously necessary this may seem, what we see, at least in practice, is more likely to be a bottom-up approach to development driven by past experience with damage, rather than a preventative top-down approach.

1.2.4 Automotive Systems Engineering

Methodology for developing systems for a vehicle, or a vehicle as a system.

Automotive Systems Engineering is simply a restriction to one implementation area up to the outermost system boundary—the vehicle as the system. However, as shown below, there are some specific features of this implementation domain which do not apply in otherwise comparable domains such as aerospace or medical technology.

1.3 Characteristics of Systems Engineering

1.3.1 System Structure, System Architecture

As noted in the definition, a system has a structure which determines the allocation of tasks, the rules governing the interaction of its elements, and the interfaces. This structure can be functionally oriented, include physical principles, or be procedural in nature.

The system architecture is a particular way of looking at the structure as designed, and accordingly there are often a number of different perspectives of the structure, which are then combined to make up the architectural model.

1.3.2 Multidisciplinary Development

Advances (particularly in electronics) are expanding both functional and non-functional possibilities so vigorously that today systems are virtually impossible to develop within one engineering discipline. As shown in Fig. 1.1, for the manifold components contained in a road vehicle, a similar manifold of disciplines is necessary.

1.3.3 Complexity

Advances in development—again most vigorously driven by the progress of electronics—have led to increasingly complex systems, which in their turn often

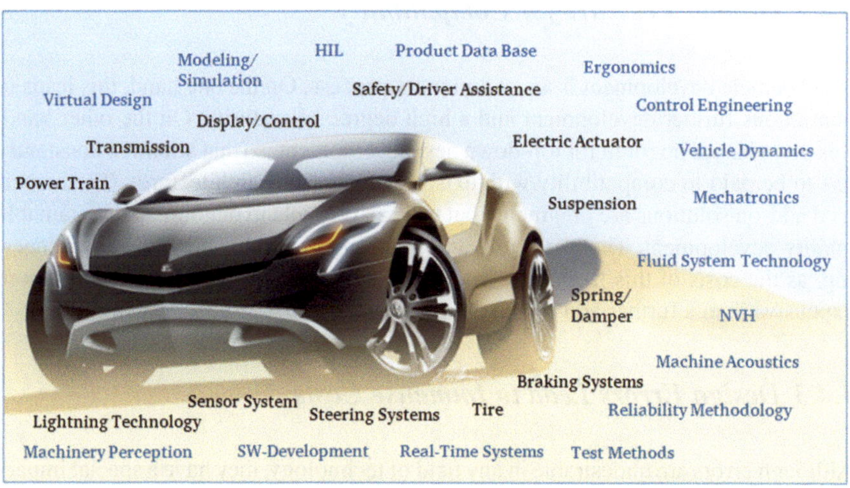

Fig. 1.1 Multidisciplinarity in Systems Engineering (Winner 2012)

have complex elements, e.g. control device hardware and software. These new possibilities in design are leading to desirable features such as expanded functionality, improved efficiency or self-diagnosis. However, this increased complexity also means that individuals can no longer retain an overview of interrelationships down to the last detail.

1.3.4 Division of Labour

Increased complexity and transdisciplinary tasks mean that a growing number of people and institutions must be involved, so that besides the technical interfaces it is also necessary to consider interfaces between individuals and between organisers.

1.4 Specifics of Automotive Systems Engineering

1.4.1 Systems Under Variable Conditions

A road vehicle is used in widely differing situations. There are different countries, often with their own regulations, different road conditions, different climatic or topological conditions. Road vehicles are driven by drivers with differing levels of skill, differing mentalities or differing driving cultures. Even if a modular system has been developed for a model range, the system boundaries cannot be regarded as homogeneous. Variations in engine and power train and every possible variant for options have to be taken into account in development.

1.4.2 Greater Pressure for Compatibility

Road vehicle development is an evolutionary process. On the one hand, this leads to continuous further development and a high degree of maturity. On the other hand, this often leaves no room for top-down new development, so that attention constantly has to be paid to compatibility with existing elements or architectures. The associated add-on solutions are pragmatic, but pose an obstacle to ongoing and sustainable quality development. There is rarely an opportunity for a fresh start or reengineering, as the costs of this fresh start would make the pilot system considerably more expensive than a further add-on solution.

1.4.3 Design Errors Lead to Immense Costs

Although errors are undesirable in any field of technology, they have a special impact in the automotive sector. Few other products have the combination of high safety criticality and large volume. A design error very quickly has a devastating effect in the field. The large number of vehicles involved mean that recalls cost a great deal of money, and the damage suffered by the quality reputation severely affects profitability. Even where errors are identified before series release, they can result in high costs, and in a worst case scenario can even delay the start of production. Design errors which are not safety critical can still be a burden because of warranty claims.

1.5 Methods of System Engineering

If problems arise, the first remedial response is to seek and implement measures to correct or limit problems. In addition to the concept of ‘bug fixing’ (taken over from software development), there was also in many cases an analysis of how it was possible for the error to occur. The next question is then how to avoid an occurrence of the same or a similar error. Frequently, this is the start of developing a method for improving the development process. Given the very wide diversity of the problems resulting from errors and also the causes of the errors themselves, there are methods which are mostly attributable to a specific range of causes.

1.5.1 Quality Process Standards

A number of quality management systems apply primarily to organisational processes. Keywords include Total Quality Management (TQM), Six Sigma, ISO/QS900x and software development quality systems based on Capability Maturity Model Integration (CMMI) or Software Process Improvement and Capability Determination (SPICE), which are also available in special versions adapted for the automotive sector. Despite the different goals and methodological structure in individual instances, they have the common goal of ensuring that the organisational requirements and procedures are in place for high development quality. Many also include concrete development methodologies, such as Quality Function Deployment (QFD) and House of Quality.

In recent years, ISO 26262 (derived from the root standard IEC 61508) has established a new standard for ensuring functional safety in automotive electric/electronic (E/E) systems which takes the entire development and product life cycle into consideration and unites various other methods.

1.5.2 Process Models

Process models such as the well-known V-Model and its variants and refinements link the sequence of steps in development in a similar way to the Waterfall Model, but for the first time assign test stages to the corresponding steps in development, so that there is a close connection between product specification and test specification.

The V-Model is also a basis for ISO 26262. This standard imposes a consistent uniform approach to the development of electric/electronic systems in vehicles, particularly with regard to using risk management methods.

1.5.3 System Modelling

System models in general terms often only exist for more recent developments. SysML (Systems Modelling Language) has evolved as a standard for these, based on Unified Modelling Language (UML).

1.5.4 Requirement Management

There are also powerful tools for requirement management, including traceability and change management, with IBM Rational DOORS apparently the most widespread.

1.5.5 Quality Analysis Methods

Methods for identifying weaknesses and analysing potential problems have been established for decades, even if they are not popular with developers, who often see them as merely time-consuming documentation. The use of Hazard Analysis and Failure Mode & Effects Analysis (FMEA) has, however, been unavoidable since the approach of ISO 26262, at the latest. Other techniques, such as Fault Tree Analysis or Event Tree Analysis, make possible further analysis of complex errors and causal relationships.

1.5.6 Conclusions on the Methods Available for Systems Engineering

Over the past twenty years, many methodological gaps have been closed. In fact, I am not aware of any gaps which would require the development of new procedural methods. However, there is a lack of continuity and integration of various methods. As all the methods demand a high training input and can only be effectively and efficiently used with a certain level of experience, there is additional effort required which is often associated with new roles in the development process, for example FMEA moderators. Not only does the internal organisational unit itself change, the new methods give rise to new service and certification organisations which have to be integrated into the development processes.

1.6 Challenges for Systems Engineering in the Automotive and Supplier Industry

The consequences of development errors, whether financial, legal or having an impact on image, create strong intrinsic interest in a development methodology which avoids errors and is efficient. Mastery of Systems Engineering is a competence which is vital for survival in competition. Many Systems Engineering tools are accordingly integrated into quality management. This is generally done centrally from top management level, which mostly leads to those at the implementing level seeing the associated measures as an additional burden, without any identifiable benefit to them. Willingness to work with these tools is accordingly very limited. For this reason, other priorities are eagerly cited, in order to avoid what are perceived as tedious tasks. In particular, engineers producing the basis for functions, such as functional software code, often are not aware that they personally could make an error with serious consequences, and that using the Systems Engineering methodology could protect them against this. However, managers also rarely create a system of incentives under which proactive Systems Engineering efforts to avoid a problem are rewarded more than 'heroic fire-fighting' when a problem does appear, or presentation of a new market-relevant feature. Even independent of such influences, people tend to see systems in a way that they can personally manage. With complex systems, however, this is dangerous if the process of development of a system splits into an arbitrary number of subsystems without any supportive planning of subsystem partitioning. The result is gaps or overlaps in responsibilities, and the flow of information is seriously impaired (inability to see the big picture). There is also a loss of awareness of what tasks other departments have to handle, and what problems they are facing, as well as what potential they have for contributing to the overall goal. In addition, while there is knowledge available of either the development process or product design, this is rarely brought together and networked in a way that creates sufficient understanding of both areas.

As noted above, there is no shortage of methods, but these can only be used efficiently and effectively with training and constant practice. In other words, successful use of the methods requires conscious acquisition and organisation of knowhow. Inevitably, this leads to an important role for specialists in order to avoid overwhelming engineers with too many methods in addition to 'normal' engineering work. No value is attached to the effort involved in implementing the methods, as this mostly has the nature of documentation, and while documentation is recognised as necessary, few engineers feel comfortable with documentation work. As a result, the effort involved in Systems Engineering is often seen as burdensome, because it is inadequately planned, because it is overlooked at the start of planning, and because it is not sufficiently familiar due to a lack of experience. Even planning the allocation of the work to the individual secondary levels is a challenge in itself.

Systems engineering makes heavy demands not only on the organisation and collaboration but also on the people working in system design. There is a need for abstraction at very many points in the development process, either in requirements analysis to identify a requirement independently of the solution, or in concept or test development in constantly asking whether the desired purpose can be achieved through the chosen approaches. While methods like Quality Function Deployment assist this process of abstraction, what we usually see in practice is the familiar approach of responding as quickly as possible to a problem without thinking about the effects on the overall system, or the criteria used in selecting the solution. The reasons for this behaviour may be the different capacity for abstraction of the individuals involved, but also the lack of practice in abstraction. A solution you have found yourself is also probably more motivating than a solution derived by a retreat to higher levels of abstraction and a systematic process, possibly with other team members. Another challenge is the uncertainty, particularly at the start of product design for new products. Here, the perspective gained from transferring knowledge from comparable innovation development is needed to avoid the extremes of walking blindly into a trap or getting bogged down in endless decision-making processes. A good management culture can be very helpful in finding the right perspective, if bold decision-making is encouraged.

1.6.1 Conclusion About the Challenges for Systems Engineering in the Automotive and Supplier Industry

While no shortcomings are apparent on the side of procedural methods, much ground still has to be made up in qualifying staff for effective and efficient Systems Engineering. There also seems scope for improvement in organisational structure where extensive scope exists for experimentation on best practices. Equally, management culture must be modified to cope for the requirements of Systems Engineering.

Besides the procedural methods, there are often shortcomings concerning the methodology of how to achieve the right output. What would be the best architecture, what are the right criteria for the assessment, how can the objectives be validated? Very often, these questions are "solved" by very pragmatic means based on the state-of-the-art. While this common approach saves money and time, it involves a latent, mostly unknown risk.

1.7 Systems Engineering in Academia

If we consider the above mentioned motivation for Automotive Systems Engineering, we could say that there is hardly any reason for Systems Engineering to play an important role in academia. It is very rare for systems to be developed to the point of market maturity, liability for quality is avoided, and gathering scientific knowledge through research does not require Systems Engineering.

In fact, autonomous Systems Engineering curricula are offered in engineer education, and there are elements of Systems Engineering theory even in basic engineering disciplines. In addition, courses include projects which include at least some Systems Engineering methods.

A university can contribute towards continuing professional education for engineers in industry by offering lifelong learning seminars on the total development process and the methods involved.

However, Systems Engineering can actually be a very useful approach in research as well. Complex systems require a systematic approach and also time for comprehensive understanding, which is mostly incompatible with the approach in industry under the usual tremendous time-to-market pressure. A fundamental review can evaluate current approaches and lead to proposals for new architectures based on criteria which are as universal as possible. Another role for academia in cooperation with industry is an independent perspective, which can be used to break down the 'not invented here' syndrome, although it can also lead to the sort of tensions consulting services encounter.

The ascending leg of the V-Model is also a valuable field of work for both university research and for developing methods for industrial use. A stringent validation philosophy which gives thought to the corresponding tests and criteria for passing them as early as the stage of formulating requirements ensures a high level of clarity right from the initial stages. Fundamental consideration of this nature can eliminate the apparent necessity of much work that people would like to be doing, but it can also avoid the vulnerability of subsequent results. It also makes it possible to create test platforms earlier for development, which in turn can be reflected earlier in the development process, e.g. through use in development and simulation environments.

The development of neutral, scientifically based evaluation criteria independently of implementation responsibility is an ideal domain for research which both satisfies academic needs and meets the needs of industry, where in-house knowhow is often insufficient. This also results in the introduction of new verification and validation methods, either as one-off experiments to obtain information or as new test systems integrated into industrial testing. Even test results are often surprising in themselves, providing new information and stimulating new investigations. Another source of demand is studies aimed at improving the efficiency and effectiveness of existing methods in design and testing.

1.7.1 Conclusion: Systems Engineering in Academia

Systems Engineering should not be regarded as an autonomous academic research discipline in parallel to mechanical engineering or control theory etc., even if the need for teaching is accepted. This is obvious in the absence of a scientific community or consistent research language.

On the other hand, Systems Engineering must accordingly be regarded more as a methodological framework for system relevant research. The goal for research methodology must be to be better than industrial Systems Engineering—in other words, to serve as a model.

However, Systems Engineering research is not a goal in itself, and must deliver substantive and usable findings. In all this there is still the challenge of gaining high academic priority for Systems Engineering competence, failing which a Systems Engineering education would lack credibility without the corresponding research activity.

Combining the traditional disciplines with a consistent Systems Engineering approach enables a high potential for progress, for both academic use and for commercial use. Designs of architectures of complex systems, objective evaluation criteria, and test methods are exemplary results of such research.

1.7.2 Examples of Automotive Systems Engineering at FZD

Modern Advanced Driver Assistance Systems (ADAS) have established a standard of driving comfort and safety unknown so far. Vehicles have become increasingly "intelligent" allowing the driver to delegate specific subtasks of vehicle guidance to these systems or to let the automation take over vehicle guidance completely in emergency situations. However, the scientifically proven advantages of ADAS are accompanied by an important disadvantage: increasing complexity. Today, most ADAS are developed separately, with the consequence that each of these systems has its own user interface and interaction concept. Indeed, the uncountable number of buttons placed all over the dashboard in a modern car almost recalls the cockpit of an aircraft. This complexity runs counter to the original goal of enhanced comfort and safety.

The automation of driving might be a solution to some of the challenges of future mobility concepts. Work on fully autonomous cars has undeniably made extensive advances in the last ten years. But fundamental challenges, for example the question of possible approval processes for road traffic, are still unsolved. On the other hand, ADAS capabilities are increasing rapidly and many have already proven themselves in real road traffic and in usage by customers. FZD's research activities are interlinked within a development roadmap, depicting a step-by-step approach from today's ADAS to fully automated driving, see Fig. 1.2. A major step towards solving the described problems of greater user complexity when combining multiple assistance systems might be "Cooperative Automation". The technical feasibility of this concept is being assessed at FZD using the "Conduct-by-Wire" approach that consists of replacing today's control elements by a single maneuver interface. This allows a maximum degree of automation, while—unlike fully automated concepts—still keeping the driver responsible for vehicle guidance in line with the Vienna Convention on Road Traffic (see Chap. 6).

The concepts for all steps between manually and fully autonomously driving are faced with the challenge of finding architectures with appropriate Human-Machine-Interfaces. This is much more than just presenting information on displays or

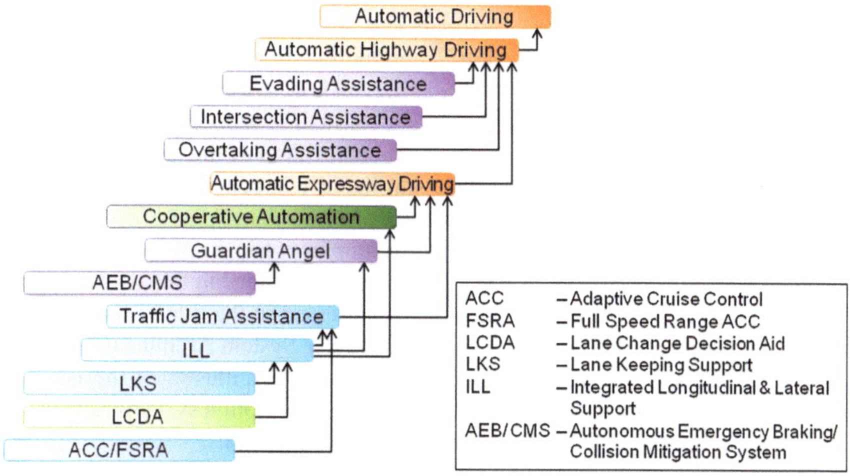

Fig. 1.2 Evolution of Driver Assistance Systems (based on: [Winner and Wolf 2009, p. 672])

providing switches for control. The interaction between the human being and the machine are multifaceted and at different levels, for example intervention at safety critical-situations, prevention of such situations by information, warning or cooperative control. To address this general problem, an exemplary concept developed in the PRORETA project, its use cases and architectural consequences are discussed in Chap. 3.

Another example where Systems Engineering opens doors to new technical solutions is the "eco2DAS" research project (see Chap. 4). The aim here is to raise the efficiency of mobility by building bridges between functionally separated islands, for example by providing the engine management with information from the ADAS environment sensors. Then, for example, the vehicle could react earlier to a traffic jam and—for instance by charging the batteries of a hybrid vehicle—decelerate the vehicle more efficiently than is the case today.

The progress in sensor technologies means that modern mechatronic systems, independently of the different purposes followed, use a manifold of signals for their task. Consequently, the functional quality, and very often the vehicle safety, depend on the quality of the signals delivered by the sensors. Hardware redundancy concepts like 2-of-3 are able to ensure a high capability of self-diagnostics, but lead to very high costs and a reduced reliability. With analytical redundancy the signal quality as well as the self-diagnostics capability can be improved. Many methods are known, but the question is to ascertain how the signal quality can be determined in order to guarantee a signal with high integrity, ensuring that at every moment the potential error is known. This is particularly challenging in the case of multi-signal fusion concepts. Chapter 10 describes an example for the fusion of satellite-based signals with sensor signals derived from an Electronic Stability Control system in order to generate a signal of high accuracy and full integrity.

Another topic relevant for the assessment of functional safety is the controllability of unintended situations due to failures and functional limitations. The above mentioned standard ISO26262 gives some rules on how to assess controllability, but it is restricted to system failures and does not consider functional limits which cannot be excluded totally when using a machine perception of the surroundings. The challenges involved in determining the controllability in general and extending the idea of ISO26262 to functional limits are described in Chap. 7.

And I would not like to fail to mention the examples of former Automotive System Engineering research projects of FZD which are not included in this book.

The system engineering aspect is directly addressed in the title of the reference (Darms and Winner 2005; Darms 2007): "A Modular System Architecture for Sensor Data Processing of ADAS Applications". This work illustrated the need for application-related processing of signals of machine perception systems as well as a concept for common, modular fusion for different applications.

The Darmstadt method EVITA for assessing the effectiveness of frontal collision countermeasures works with real systems in realistic situations for the purpose of avoiding frontal collisions (Hoffmann and Winner 2007, 2009; Hoffmann 2008). In this way, totally different countermeasures from warning elements to active brake intervention could be assessed (Fecher et al. 2008). As part of the functional system research on new criteria for evaluation and dimensioning, Hohm derived and determined the Adequate Overtaking Margin criteria (Hohm 2010) for dimensioning an overtaking assistance system first presented by Hohm et al. (2008). Habenicht (2012) was able to determine the effectiveness of his new maneuver-based lane change assistance system (Habenicht et al. 2011) by introducing an Available-Time-to-React criterion as measure for driving safety and conducted experiments to compare different approaches for lane change assistance. In the same general way, Lattke (2012) and Lattke et al. (2011a,b) evaluated a Cooperative Driver Assistance System for Motorcycles, coming from the assistance goal to assessment criteria and comparative experiments to determine effectiveness.

1.8 Conclusion

Systems Engineering in general and Automotive Systems Engineering in particular are key competence fields to cope with complex systems in the automotive world, no matter whether in industry or in academia. In industry the ability to follow the Systems Engineering approach is essential for future success. In the academic world, Systems Engineering as a methodological framework means avoiding research arbitrariness. It links the research topic to the desired overall goals, makes it objectively verifiable, and generates development and verifying methods and tools for practical use in industrial development. It transfers a deep comprehension about the system to innovative solutions.

References

Darms, M., Winner, H.: A modular system architecture for sensor data processing of ADAS applications. In: IEEE Intelligent Vehicles Symposium, Las Vegas, USA, 06–08 June 2005

Darms, M.: Eine Basis-Systemarchitektur zur Sensordatenfusion von Umfeldsensoren für Fahrerassistenzsysteme. Dissertation Technische Universität Darmstadt, Fortschritt-Berichte VDI, Reihe 12, Nr. 65, Düsseldorf (2007)

Fecher, N., Fuchs, K., Hoffmann, J., Abendroth, B., Bruder, R., Winner, H.: Analysis of the driver behavior in autonomous emergency hazard braking situations. In: FISITA World Automotive Congress, München, 14–19 Sept 2008

Habenicht, S.: Entwicklung und Evaluation eines manöverbasierten Fahrstreifenwechselassistenten. Dissertation Technische Universität Darmstadt, Fortschritt-Berichte VDI, Reihe 12, Nr. 756, VDI Verlag GmbH, Düsseldorf (2012)

Habenicht, S., Winner, H., Bone, S., Sasse, F., Korzenietz, P.: A maneuver-based lane change assistance system. In: 2011 IEEE Intelligent Vehicles Symposium, Baden-Baden, 05–09 June 2011

Hoffmann, J.: Das Darmstädter Verfahren (EVITA) zum Testen und Bewerten von Frontalkollisionsgegenmaßnahmen. Dissertation Technische Universität Darmstadt, Fortschritt-Berichte VDI, Reihe 12, Nr. 693, Düsseldorf (2008)

Hoffmann, J., Winner, H.: Das Darmstädter Dummy Target EVITA—Ein Werkzeug zur Beurteilung von Antikollisionssystemen, VDI-Tagung Erprobung und Simulation (2007)

Hoffmann, J., Winner, H.: EVITA—Das Prüfverfahren zur Beurteilung von Antikollisionssystemen. In: Winner, H., Hakuli, S., Wolf, G. (ed. 2012) Handbuch Fahrerassistenzsysteme, pp. 69–75. Vieweg + Teubner Verlag, Wiesbaden (2009)

Hohm, A., Wojek, C., Schiele, B., Winner, H.: Multi-level sensorfusion and computer-vision algorithms within a driver assistance system for avoiding overtaking accidents. In: FISITA World Automotive Congress, Munich, 14–19 Sept 2008

Hohm, A.: Umfeldklassifikation und Identifikation von Überholzielen für ein Überholassistenzsystem. Dissertation Technische Universität Darmstadt, Fortschritt-Berichte VDI, Reihe 12, Nr. 727, Düsseldorf (2010)

Lattke, B.: Ein kommunikationsbasiertes Gefahrstellenwarnsystem für Motorräder. Dissertation Technische Universität Darmstadt, Fortschritt-Berichte VDI, Reihe 12, Nr. 757, VDI Verlag GmbH, Düsseldorf (2012)

Lattke, B., Sperber, F., Müller, T., Winner, H., Eberlein, R., Hoffman, R.: MoLife—hazard detection in a cooperative assistance system for motorcycles. In: Proceedings of the 22nd International Technical Conference on the Enhanced Safety of Vehicles, Washington (2011a)

Lattke, B., Hanßmann, D., Müller, T., Winner, H., Eberlein, R., Hoffman, R.: Basic science for a new cooperative driver assistance system for motorcycles (MoLife). In: Proceedings of the First International Symposium on Future Active Safety Technology Toward Zero-Traffic-Accident, Tokyo (2011b)

Winner, H, Wolf, G.: Quo vadis, FAS? In: Winner, H., Hakuli, S., Wolf, G. (ed.) Handbuch Fahrerassistenzsysteme. Vieweg + Teubner, Wiesbaden (2009)

Winner, H.: Challenges for automotive engineering. Public Serv. Rev.: Eur. Sci. Technol. **14**, S230–S231 (2012)

Chapter 2
Automotive Systems Engineering: A Personal Perspective

Markus Maurer

2.1 Motivation: Do We Need Yet Another Theoretical Framework?

The complexity in the development of automobiles grows. In the 1960ies three body variants (sedan, sports car, and spider) were sufficient to satisfy customer needs (Heißing and Ersoy 2008). After 2000, Heißing distinguished 16 different types of variants (Heißing and Ersoy 2008).

The portion of software and electronics at the added value is rising significantly. It changes the requirements on the development process itself. In addition, other tools and methods in car development are becoming necessary.

The concept of automotive systems engineering sketched in this article is shaped by my personal experience as a researcher and developer in the field of autonomous road vehicles and driver assistance systems.[1] In the late 1980s and the early 1990s researchers packed more computers into road vehicles than a sedan could carry. It was a milestone when sedans were able to demonstrate automatic driving functions in public traffic for the first time (Fig. 2.1, Maurer et al. 1994).

In 2007 autonomous vehicles demonstrated in the framework of the DARPA Urban Grand Challenge that they were able to complete missions and, for a limited time, survive unmanned in a cooperative environment on a shut-down military airport among other autonomous vehicles and carefully operating human drivers (Darpa 2008).

Driver assistance systems require complex software systems in order to perceive the environment and to execute appropriate actions supporting the goals of the

[1] In this article systems are called driver assistance systems if they recognize their environment and the current situation with machine perception in order to support the driver appropriately in his driving task (Maurer 2012a, b).

M. Maurer (✉)
Institute of Control Engineering, Technische Universität Braunschweig,
Hans-Sommer-Str. 66, D-38106 Braunschweig, Germany
e-mail: maurer@ifr.ing.tu-bs.de

M. Maurer and H. Winner (eds.), *Automotive Systems Engineering*,
DOI: 10.1007/978-3-642-36455-6_2, © Springer-Verlag Berlin Heidelberg 2013

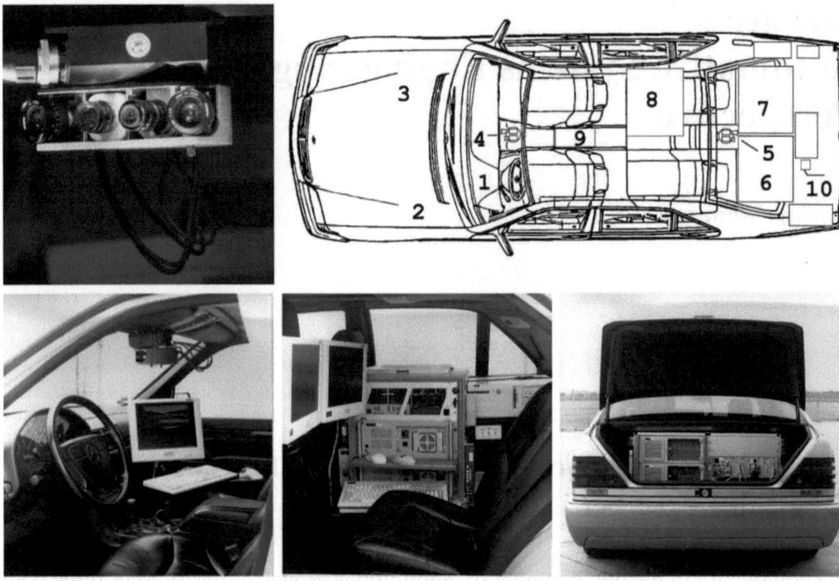

Fig. 2.1 VaMP: Automated driving functions with computer vision in the 1990s (Maurer 2000a, 2000b)

drivers. After two decades of personal experience, joint projects, and interviews with colleagues from Audi, BMW, Bosch, Continental, Daimler, Toyota, and Volkswagen, I can conclude that it is still a major challenge to develop complex, distributed safety-critical, mechatronic systems on time and in accordance to the quality and financial goals of the projects.

Lately the reports of developers of hybrid vehicles from the companies above resemble those of the developers of driver assistance systems. Here too, the focus is on complex distributed safety-critical systems; again innovations are created which deeply affect the architecture of the overall vehicle.

This article tries to elucidate why it is so challenging to develop driver assistance systems or hybrid drive train systems. An analysis by means of systems engineering can help to create tools and methods for vehicle development and to adapt processes to the changing requirements.

2.2 Perspectives in Vehicle Development

Historically the development departments of the German car manufacturers have known two job profiles: the design engineer (in German: Konstrukteur) and the application engineer (in German: Versuchsingenieur). The design engineer develops a component according to the defined requirements, in the early days at the drawing board, nowadays at his CAD-workstation. The design work itself is often outsourced to engineering offices. In this case, the design engineer of the car manufacturer controls external developers at their CAD-workstations.

The application engineer proves the component in the context of the vehicle. He runs test beds and tests the component in operation, drives it under all possible conditions. It is apparent that both job profiles are oriented towards the **component**.

At Audi the job profile of the application engineers has been enhanced to that of the property engineers over the last years (by example Kohoutek et al. 2007).[2] They are responsible for properties of the vehicles which can be determined by multiple components. For instance, the property braking distance is not only determined by the wheel brake itself but by the axles, the tires and the application of several active control systems.

Horst Glaser, head of the Audi chassis development department for many years, underlines the importance of property engineering for the development of a modern chassis: The chassis development department stands for the customer values as safety (active safety), driving dynamics, driving pleasure, driving comfort... The property engineers assess the properties of existing vehicles and for future vehicles due to customer relevant criteria such as agility, directional stability, braking, steering behavior, riding comfort, cornering, traction, safety... In many years of training they have gained a subjective sense of driving based on their haptic perception. They are familiar with customer expectations and can describe them on a subjective scale. Their work is supported by measurement and development tools specified by them. All of these skills in chassis engineering have grown over decades (discussions for Glaser 2006). Based on their experience and in cooperation with the design engineers and system partners they develop competitive chassis properties.

Up to now, two important perspectives have been established: The perspective of the component oriented design engineer and the perspective of the customer oriented property engineer.

In this article I claim that an additional systemic perspective is needed (Fig. 2.2). Whether or how this will change the job profiles discussed above is of no significance. But it is essential to add the systemic perspective to those already established.

2.2.1 Definition: Capabilities of a Vehicle

For the sake of academic habits a short definition of properties is given. Readers more interested in "the big picture" can skip the next few lines.

Properties are an abstraction of the vehicle and its components. They emerge from the interaction between components. Properties of a vehicle are modeled appropriately if the simulated model shows a behavior which is sufficiently similar to the behavior of the vehicle itself according to given metrics.

[2] Up to now we haven't found an established English translation for the German "Eigenschafts-entwickler". The direct translation "property developer" has an established different meaning. Therefore we use the term "property engineer" in this text.

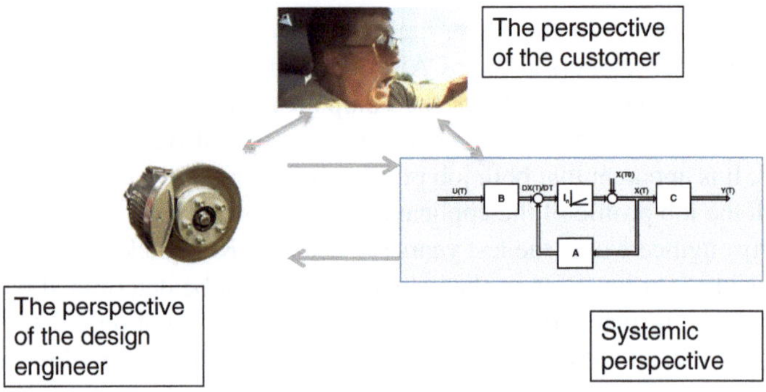

Fig. 2.2 Perspectives in vehicle development [picture of brakes: Bergmiller, Personal Communication, Braunschweig, picture of driver (Färber and Maurer 2005)]

System theory supplies a quantitative description language for modeling vehicle properties. In automotive systems engineering the vehicle is regarded as a system in order to describe the properties of the vehicle quantitatively and to develop them systematically.

2.3 Potential of Automotive Systems Engineering

A paradigm is of any practical meaning if its introduction promises significant use during practical work. The consistent use of systems engineering in the vehicle development supplies methods appropriate to describe the properties of the vehicle objectively. Today it is good practice in many fields to evaluate the properties subjectively. Often there is a lack of appropriate objective metrics and reference tools.

Mastering the complexity inherent to the development and production processes is a core competence of modern car manufacturers. Whereas subsystems are developed, produced and supplied by system suppliers in many cases, the car manufacturers can only succeed if they are able to develop entire vehicles on time meeting quality, financial and market goals. Systems engineering offers rich methods supporting mastering of complex systems: Manifold architectures, methods for simulating and testing vehicle properties during the development phase and methods for developing processes belong to the toolkit of systems engineers.

The systematic consideration of functional safety is of growing importance as distributed mechatronic systems become more and more responsible for safety-critical properties. The automotive standard ISO 26262 replacing the industrial standard IEC 61508 underlines the meaning of functional safety in vehicles in the development process.

The three items introduced above

- objective evaluation of properties,
- methods for mastering complexity,
- functional safety

Fig. 2.3 City pilot: Autonomous Driving on the central ring road in Braunschweig (Wille et al. 2010)

will help to structure this article. Examples given below and throughout this book will illustrate the practical impact of these concepts.

2.4 Objective Evaluation of Properties

In the framework of the project city pilot (in German: "Stadtpilot") in which our group investigates methods for autonomous driving in the inner city, both functional and methodical goals were defined. The latter goals include the development of goal measures and metrics for machine perception and machine behavior decision (Fig. 2.3, Wille et al. 2010).

A practical example will illustrate the meaning of metrics for machine perception: The current Audi A8 is equipped with various sensors for environmental perception (Fig. 2.4): Two 77 GHz radar sensors perceive the vehicles in front of the ego-vehicle and enable it to keep a safe distance to the vehicle ahead ("Adaptive Cruise Control", ACC, (ISO 2002)) or to realize predictive safety systems ("Forward Collision Warning and Avoidance", (Maurer 2012a)). A video camera allows perceiving the lanes

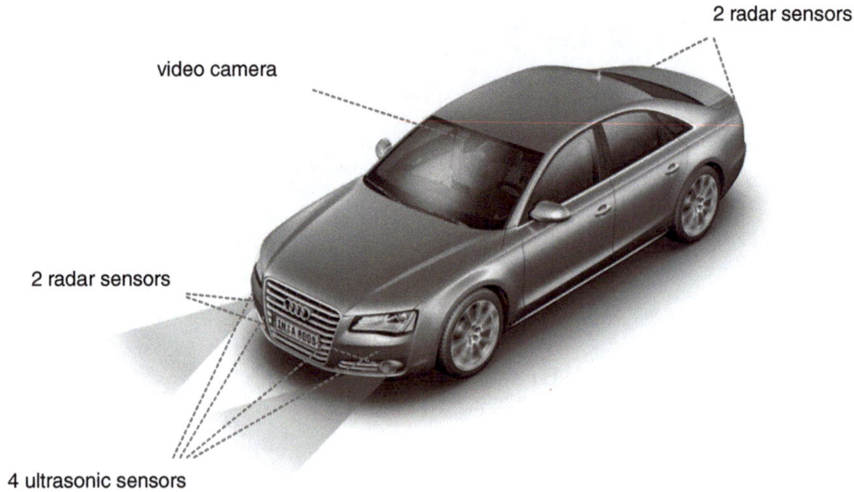

Fig. 2.4 Audi A8, model year 2010: Sensors for environment perception (Duba G-P, 2010, Ingolstadt, personal communication)

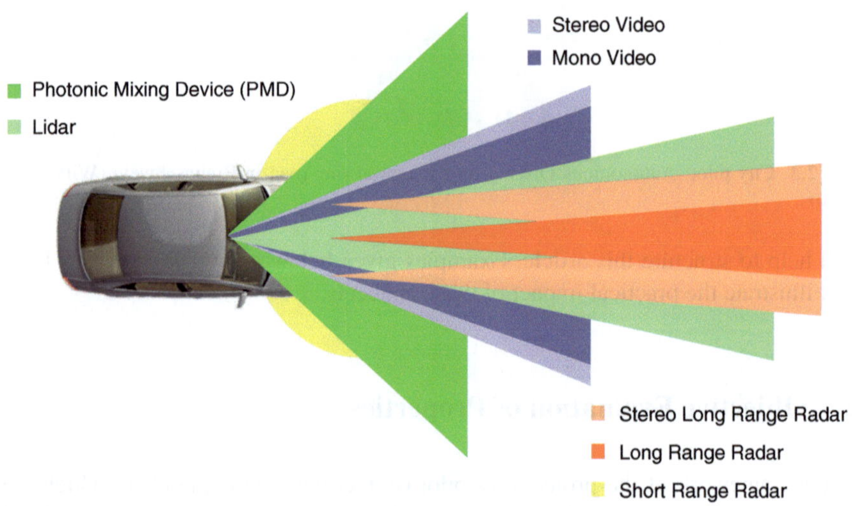

Fig. 2.5 Potential sensors for the look-ahead range of the current Audi A8 (Maurer 2007)

and the traffic signs and redundantly recognizes leading vehicles. Ultrasonic sensors support the customers when parking also when looking forward.

How does the selection of sensors work? Figure. 2.5 illustrates the forward looking sensors examined during development phase of this vehicle. In addition to the sensors finally used in mass production a variety of sensors was evaluated: PMD (Photonic Mixing Device), lidar, mono radar-, short range-radar and stereo video sensors, almost each type from several suppliers. Due to the lack of established metrics and reference sensors, the choice was based on internally defined criteria and on expert knowledge. The development of new sensors could be more goal-oriented if next generation sensor requirements could be derived from the requirements of future assistance functions. The required sensor performance should be quantitatively defined in established metrics testable with appropriate reference sensors.

2.4.1 Test Bed Reference Sensors

Funded by the German Research Foundation (in German: "Deutsche Forschungsgemeinschaft", DFG), a test bed for reference sensing will be further developed by our group. It will supply the metrological base for the development of metrics. With this test bed, we pursue the following goals:

- Establishing "ground truth" with the required accuracy,
- Establishing metrics generally accepted,
- Establishing benchmarks generally accepted for comparing sensors even depending on the current mounting place,
- Establishing reference sensors as standard tools in all internal projects for automotive machine perception,
- Development of sensor models.

The test bed consists of a carrier vehicle and a 3D-laser as stationary reference sensor. The vehicle will be equipped with precise inertial-based ego-positioning coupled with DGPS. The environment can be perceived with radar, lidar and camera sensors which can be mounted on different positions. In addition, a 360° sensor from Velodyne and a 360° camera sensor are available. All sensor data are tagged with timely synchronized time stamps. Ground truth within the scope of required accuracy will be created by fusing the sensor data either online or in stationary scenarios offline (Fig. 2.6).

Brahmi (2013) discusses requirements on reference systems, introduces metrology definitions helpful for referencing signals and reports experimental results with an INS/DGPS based reference system and its limitations.

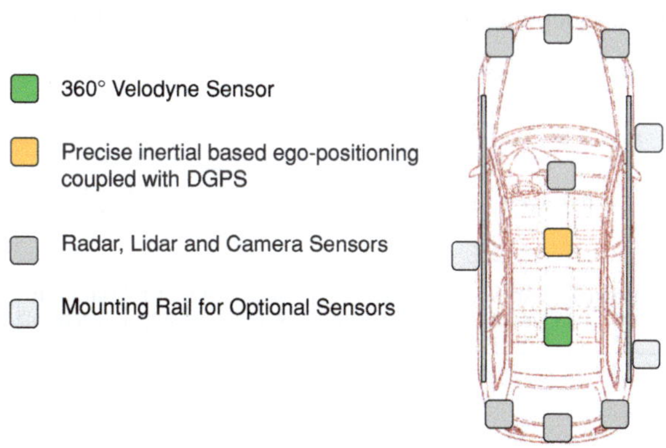

Fig. 2.6 Test Bed: Reference Sensors

2.5 Mastering Complexity

The growing complexity in vehicle development is a major motivation for the paradigm of automotive systems engineering (see above). Methods of systems engineering will help to master the complexity in the vehicle. The following methods for managing complex systems will be distinguished in this article:

- Architectures,
- Methods and tools for simulation and test of vehicle properties in the development phase, and
- Processes for system design.

2.5.1 Architectures in Vehicle Development

In this article, hardware-independent architectures are discriminated from hardware-specific architectures (Fig. 2.7).

From a historic point of view, hardware-specific architectures dominated due to the component orientation of the car manufacturers. The hardware architectures, also called technical architectures, (e.g. Reif 2006) define the relative position and motion of the components.

The term hardware architecture includes network topologies of the embedded computers connected by data busses as well. In this sense, Fig. 2.8 illustrates the data bus topology of the current Audi A8 as an example for hardware architecture.

Traditionally, software of an ECU (electronic control unit) was tailored and optimized for this particular embedded computer. This kind of hardware orientation is removed by introducing AUTOSAR (2012); software and software architectures are becoming increasingly hardware-independent. Figure 2.7 takes this development into account: There is a distinction between hardware-specific and hardware-independent aspects of the software architecture.

By definition hardware-independent architectures benefit from the advantage that they follow criteria independent of the actual hardware target platform. Functional

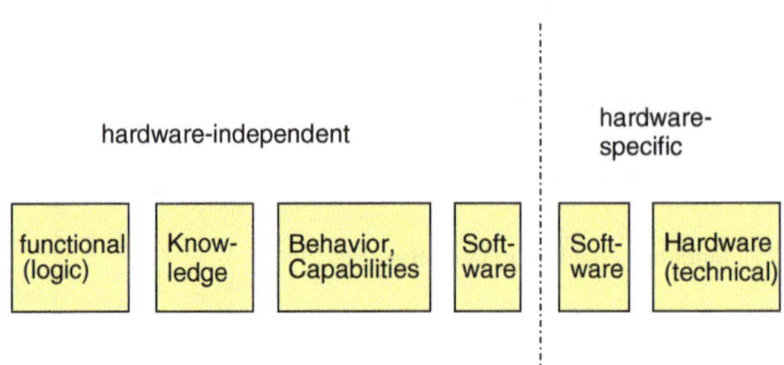

Fig. 2.7 Architectures in Vehicle Development (Maurer 2012c)

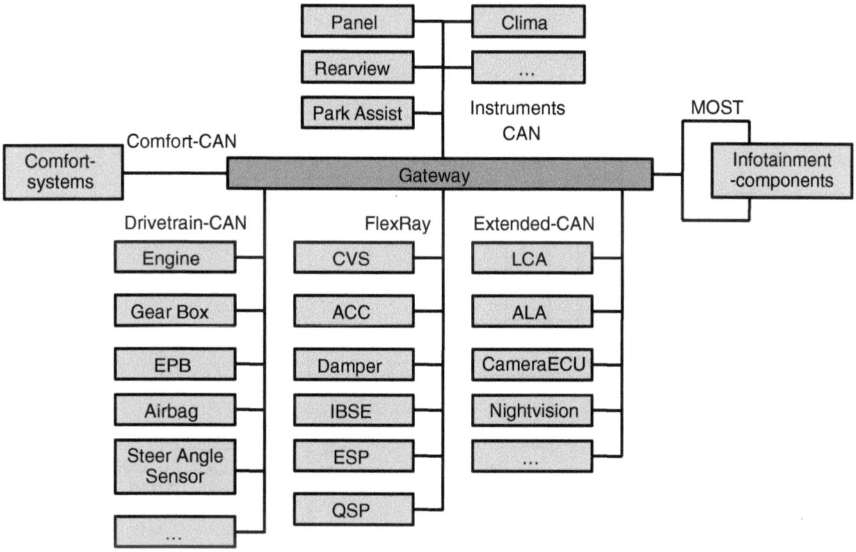

Fig. 2.8 Electronic hardware architecture of the current Audi A8 (Kötz J, Ingolstadt, Personal Communication). *EPB* electrical parking brake, *CVS* computer vision system, *ACC* adaptive cruise control, *IBSE* inertial based state estimation, *ESP* electronic stability program, *QSP* Quattro sport, *LCA* lane change assist, *ALA* adaptive light assist

aspects, the (central) knowledge representation in the system and the vehicle properties are recommended as classification criteria for hardware-independent architectures in this article.

With the evolving job profile of the application engineer, car manufacturers began to define topologies for vehicle properties, which increasingly lead vehicle development (Kohoutek et al. 2007).

The functional system architecture is of outstanding importance for the sustainable development of complex systems. Unfortunately, we cannot discuss a functional system architecture employed by a car manufacturer due to non disclosure agreements. But even the discussion of a functional system architecture developed during my own PhD thesis at the University of the Armed Forces Munich in the 1990s shows the value of this way of system description. In Fig. 2.9 this functional system architecture is sketched. A detailed discussion is given in Maurer (2000a) (in German) and a compact version is available in English (Maurer 2000b). The interesting aspect emphasizing the meaning of functional system architecture is that these architectures can be valid for a long time. Compared to recent architectures as described in Lotz (2013) or Geyer (2013) only extensions have been added which easily fit into the aged version. In the same time span the hardware architectures were revolutionized by the introduction of several data busses and gateways.

Nowadays the aspect of knowledge representation is still underrepresented in modern vehicles: Today important system knowledge is frequently represented

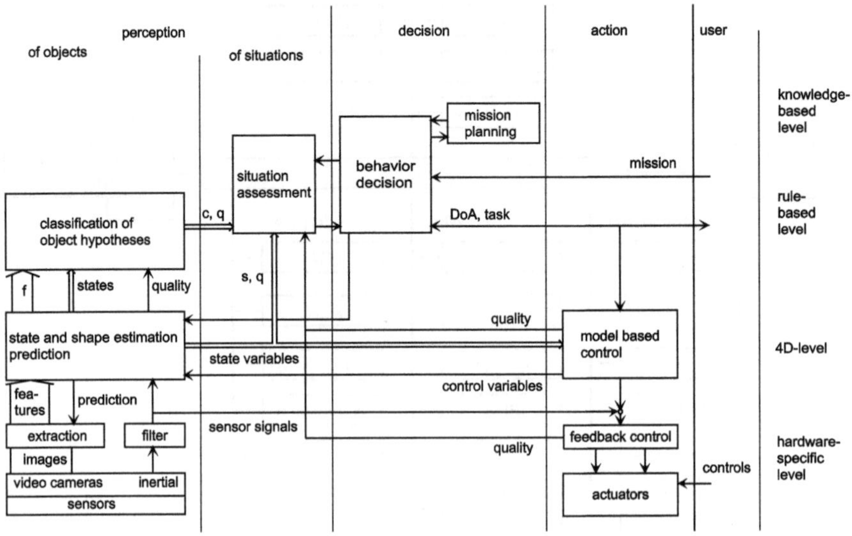

Fig. 2.9 Historic Functional System Architecture (Maurer 2000b). *f* features, *c* class, *q* quality, *s* state variables, *DoA* degree of automation

implicitly in diagnosis functions of distributed ECUs. In the future the explicit central representation of the ego-vehicle and its capabilities will become an important prerequisite for operating highly automated vehicles. A vehicle with increased autonomous capabilities needs to know these capabilities in order to recognize which of them can safely be executed automatically.

The difference between a decentralized implicit representation and a central explicit representation is illustrated by the following example. Adaptive cruise control systems monitor whether a radar sensor can perceive undisturbed or whether it will become "blind", unable to recognize relevant objects any longer. The latter problem can occur if it is raining with droplets of a certain size or if the radome is soiled. In this case, the system has to recognize that the sensor is going "blind"; the ACC function is to be deactivated. The automation of ACC will not be offered to the driver until the perception situation has improved again.

In an explicit representation classes for the radar and further advanced classes for the perception capabilities of a vehicle will be defined. The possible degree of automation represented in another class will depend on the current perception capabilities. The representation of the degree of automation and even the perception capabilities will be available both locally on the ECU ACC is placed on and centrally for a central decision unit responsible for the overall vehicle behavior.

In an implicit representation the blindness of the radar can lead to a state transition in an automaton from a state "ACC active" to a state "ACC inactive" or "Sensor blind". The objects for the sensor itself, for the capabilities and for the degree of automation will not be established in the system. The knowledge will be kept locally.

The possible extension of the latter traditional approach is more difficult: Other vehicle states and events will lead to additional states and additional transitions until the complexity of the automata or the statecharts are difficult to handle. It is obvious that the form of local treatment of diagnosis modes can reach its limits too. Dependent on the driver type, it will become obscure for the driver why a function currently is not available. The explicit represented capabilities of the object-oriented approach are a good basis to explain system behavior to the driver.

In highly automated civil automobiles an explicit central knowledge representation will be needed depending on the system complexity. A way of central representation simplified for an autonomous research vehicle was realized in the 1990s (Fig. 2.10, Maurer 2000b). At several levels the vehicle state is represented centrally. Dependent on the "health state" a central decision unit determines which functions can be automated and which must not. An interface is to be established between the central behavior decision and the human driver or supervisor explaining the system state to the user both compact and as self-explaining as possible.

This shows also a trend for the development of onboard-diagnosis. Methodologically, object-oriented programming will support this trend. If possible the classes will be independent of their target hardware. This development will be simplified if

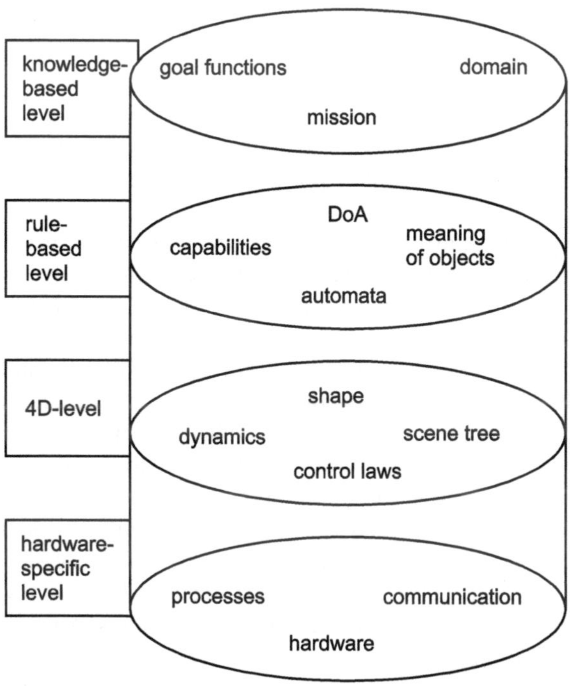

Fig. 2.10 Self-representation of an autonomous vehicle (Maurer 2000b). *DoA* degree of automation

the number of ECUs decreases as predicted and the computation power required is supplied by a few powerful ECUs connected by powerful data busses.

As an example for a modern software architecture Ohl discusses an object-oriented architecture for sensor data fusion developed in the "Stadtpilot" project (Ohl 2013).

2.6 Methods and Tools for Simulation and Test

The growing number of automobile variants mentioned initially and the growing complexity connected with their development require new methods and tools for simulation and test. Innovative tools can significantly speed up the development time and reduce development costs especially if big development steps between two successive models are due. The introduction of hybrid and electric drive trains supplies many examples for challenging development steps. Appropriate tools can help to reduce the number of prototypes needed in the development process. Some innovative methods cannot be tested without new tools due to safety reasons.

In this article an example for the latter motivation will be sketched. When developing an automatic emergency brake the question arose how the reaction of human drivers to an automatic emergency brake could be tested appropriately. Therefore different tools were developed. The most innovative will be introduced shortly.

The emergency brake should be developed in a way that an emergency braking is to be triggered once an accident cannot be avoided any longer within the limits of handling (Kopischke 2000). This way a system for collision mitigation (Maurer 2012a) was defined: A justified trigger required a collision at the end or during the braking, so that testing with real test vehicles and prototypes was only possible in justified exceptions (by example in a crash test).

Unfortunately, the testing of this function exceeds the capabilities of conventional driving simulators. They cannot realize the deceleration required combined with a natural optical impression for the human test driver. The kinesthetic feedback for the driver is not sufficient, complex visual information lack. The driver can become motion sick. The experiment identifies the limits of the simulators. It is not appropriate to evaluate the innovative function automatic emergency braking.

A vehicle-in-the-Loop-Simulator (VIL-Simulator) overcomes the limitations in traditional simulation mentioned above. Thereto a real driver rides a real vehicle on a real test track. Only the other vehicles on the track are simulated and visualized for the driver with transparent glasses for augmented reality (Bock et al. 2007). So VIL combines advantages of classic simulation like repeatability, safety, conserving resources with the advantages of testing in a test vehicle. The test rides are performed with real driving dynamics in a real environment; motion sickness does not occur with transparent glasses.

Two other examples for innovative tools for the development and test of complex systems are described in this volume. Bergmiller (2013) designed and realized a generic prototype for the investigation of novel automotive systems. Eltaher and Maurer (2013) investigated a new approach for testing multi-media systems which is based on the experience of senior application engineers. We believe that the research community should actively support the automotive development teams by designing tools for test and development.

2.7 Processes and Organizational Structure for System Design

The analysis of the perspectives has revealed that from a historic point of view the component oriented perspective of the design engineer and the application engineer have dominated in the vehicle development departments. Therefore, it is consistent that even the processes and the organizational structures of the companies were component-oriented historically. The growing complexity mentioned above also creates challenges for the process and the organizational development.

2.7.1 Processes for the Design of Complex, Mechatronic Systems

An important milestone in process development at car manufacturers was the introduction of the V-model. The V-model was originally developed for the design of complex, mechatronic defense systems. The V-model supports different fundamental design principles helping to structure complex systems. First of all, it supports the top-down design from overall requirements on the system level stepwise down to the detailed requirements on the component level.

Specifying appropriate test cases for each requirement is essential to the V-model. Thus, the top-down structure of the requirements corresponds to a bottom-up structure of the test cases (Fig. 2.11).

The introduction of the V-model as a paradigm in the design process of electronic vehicle systems leads to a significantly more structured way of development with car manufacturers and their system partners (e.g., Breu et al. 2007). The more the requirements are specified in detail, the more obvious the limited test coverage of complex assistance systems becomes.

It is discussed critically in scientific publications that the V-model may not be appropriate if the information base is not yet complete at the beginning of the design process and therefore the system cannot be developed top-down (e.g., Reif 2006). In reality, the design proceeds incrementally and iteratively; many steps of the V-model or even the whole V-model are processed several times (Schäuffele and Zurawka 2006).

A simple design model, which was developed during the research project "Automatic Emergency Braking" at Audi, takes the need for iterative design loops into account (Maurer and Wörsdörfer 2002). The process was visualized in a simple diagram: Fig. 2.12 shows a full circle containing a complete iteration loop. An abbreviation path is defined after less than half of the circle leading back to the starting point

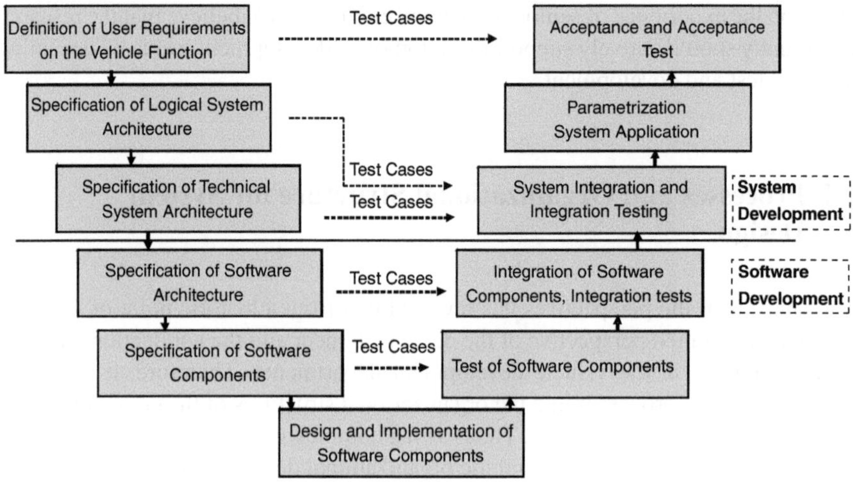

Fig. 2.11 Basic structure of the V-model

of the development process. A more technical form of the notation was presented in 2006, but not continued during the last years (Glaser 2006).

As a result, two iterative loops emerge from the structure described above: The first loop is much shorter and saves resources; it requires expert knowledge from different departments. The tasks are performed theoretically or supported by a chain of concatenated model-, software, and/or vehicle-in-the-loop tools (Bock 2009) in more advanced labs; no prototypes are built up during the inner iterative loop. The approach is extremely powerful if the experts available within the car manufacturer—supported by external experts if necessary—identify the basic design conflicts within the inner iterative loop; if, in addition, they profoundly distinguish between realizable and desirable, but non-realizable assistance functionalities.

Prototypic systems will only be built up if the experts agree to a function definition as a preliminary result solving all design conflicts revealed during the theoretical

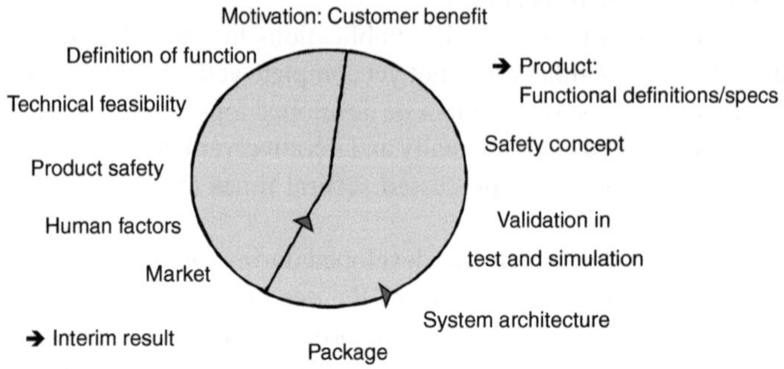

Fig. 2.12 Systematic Design of Driver Assistance Systems (Maurer and Wörsdörfer 2002)

discussion. Sometimes experimental setups may be required even in the iterative loop to solve basic questions.

The needs of the driver and the assistance to him are always the starting point of the design process. Note that for the commercial success of the system the subjective driver needs, not the objective impact of the system can make the difference.

Based on these ideas, possible assistance functions are derived and tested by the expert to answer the following questions: Can they be realized with state-of-the-art technology? Can the functional gaps and the system failures be controlled by untrained users in each situation? Is a user-transparent design of the assistance function and its limitations possible? Are there any sensible human–machine interfaces? Can the system be designed financially affordable for the customer? Does it fit the branding of the car manufacturer? A more detailed discussion including a practical example and a full iterative loop is published in Maurer (2012a).

The approach described above supplements the class of design processes collected in the field of integrated product development (e.g., Ehrlenspiel 2007). This design scheme should be taken into account in any research and development phase of a system. User-centered and holistic design should be mandatory in academic research. In the phase of industrial research and predevelopment, the design processes are important for the commercial success of the manufacturer. The fine adjustment is performed during the series development phase especially if innovative machine perception is involved; prototypes of the sensors only available shortly before market introduction reveal whether the specifications gathered at the beginning of the project will be met by real production type sensors. If they do not fulfill the specifications, it may become necessary to adjust the functionality shortly before start of production by adding yet another loop in the design scheme.

Of course, there should be open research and predevelopment projects, not directly addressed to particular user needs. But it is important that these projects are declared accordingly and do not suggest specific customer benefits.

2.7.2 Product Development Process

In automobile industry the design processes sketched above (V-model, systematic design process) are applied to the design and the development of subsystems. It would be interesting to see to which extent the principles of the V-model and iteration are reflected in the central product development processes of the car manufacturers. Are top-down-design, systematic tests, iterations, customer-oriented design supported by the central product development processes as well or are conflicts pre-programmed between the central development planning and the distributed system development?

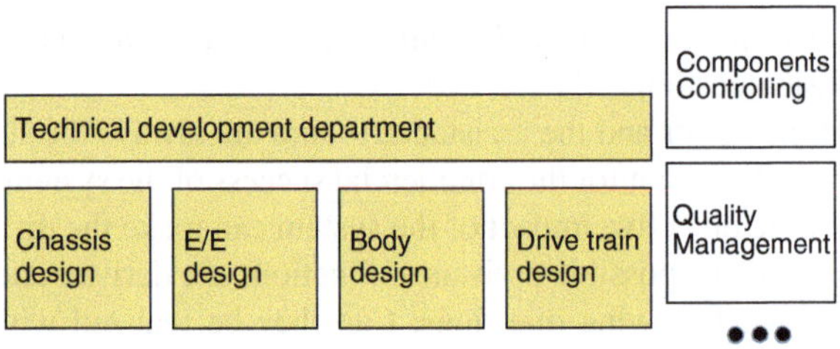

Fig. 2.13 Part of a historic organization chart of a car manufacturer

2.7.3 Organizational Structures

Also the organizational structures of car manufacturers were originally component-oriented—consistent to the component oriented job profiles and processes (see above). Figure 2.13 shows a part of a classic organization chart of a car manufacturer. The technical development department in many companies is still structured by the dominant group of components in the automobile. Classification criteria can be the chassis, drive train, interior, vehicle body, vehicle electric and electronics (E/E) or other dominant group of components. Controlling and quality management were historically built up component-oriented.

The question today is to which extent the organizations have been adapted to the requirements of distributed, complex, mechatronic systems. Are project managers of these systems supported by the organization of the company and its processes when developing properties on time according to the financial and quality goals? Or does it depend on the engagement of individuals whether property development succeeds in the vehicle project?

The significance of this question will be illustrated by an example from the field of development of driver assistance systems. Early driver assistance systems were designed by the developers of the actuators affected. Open minded engineers realized when monitoring state of art that "their" actuators could be enabled to execute actions in a new context with innovative sensors: New innovative, customer relevant functions were born. The brakes of the vehicle can decelerate the car in order to avoid an impending accident even without driver intervention if the situation is recognized by a radar system. A display will show speed limits if they have been recognized by a vision system before. The headlight will be controlled in pan and tilt direction if environmental information is available.

In different parts of the technical development department, several teams for driver assistance systems emerged as a result of this way of actuator centered development. It does not come as a surprise that these teams developed systems centered on their actuators of interest. When sensors of diverse subsystems began to monitor a similar field of view, the need for reorganization was recognized. Teams were created which developed the environmental perception for the whole vehicle.

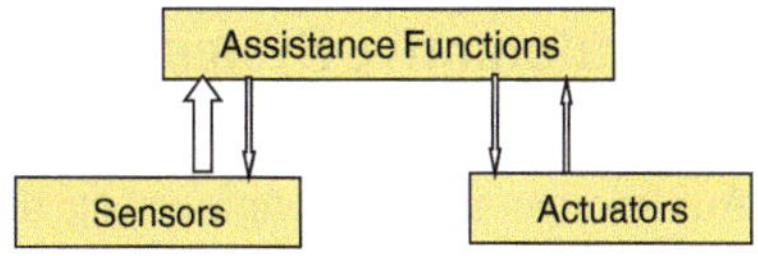

Fig. 2.14 Simplified functional system architecture of driver assistance systems

An analysis with methods from systems engineering shows that a further reorganization will yield additional benefits. The analysis of the interfaces in a simplified functional system architecture shows that the interface between function and actuator can be described straightforward (right side in Fig. 2.14). The interface between perception and the function is much more complex and less understood. This interface will be extended, optimized and modified many times over the coming decades (left side in Fig. 2.14).

The interfaces between units in an organization will be defined in such a way that they are easy to describe and well understood. This means that the interfaces between the development team of driver assistance systems and other organizational units should be defined between the teams for actuator development and the teams for the development of assistance functions. The interfaces will be hard to handle if the development teams for driver assistance systems are kept distributed in the component oriented structures of the manufacturer with the major interface between a central machine perception unit and decentralized driver assistance development teams. This example underlines that even a simple systemic analysis reveals potential for economically relevant organizational optimization.

2.8 Functional Safety

The original conception of our team did not contain functional safety as an explicit topic of research. In the meantime we have learnt from our experiments with several experimental vehicles in diverse projects that functional safety significantly influences architectures, test and development methods and process development. In addition, innovative mechatronic subsystems offer many technical possibilities directly linked to unsolved questions of functional safety. This is why we decided to focus also on functional safety aspects.

In the project city pilot mentioned above, a PhD thesis is under development devoted to safety critical aspects of the system (Reschka et al. 2012). In the current research phase a safety driver and an additional operator represent the safe state when driving highly automated.

In Bergmiller (2013) a vehicle is presented that requires numerous innovative safety concepts to be driven only on dedicated test tracks by trained drivers.

2.9 Automotive Systems Engineering in Research and Teaching

Automotive systems engineering describes a way of thinking demanding time to study. Ideally, it is already taught at university. This is why TU Braunschweig has established a professorship for Automotive Electronics Systems in addition to the classic component oriented education. According to the methods to be applied this chair is placed at the Institute of Control Engineering. Derived from the practical research of the group partly described in this article and in this book, a theory of automotive systems engineering is being developed. It will supply terminology and methods for the systematic development of vehicles. The lecture series dedicated to this topic is held for one semester containing two hours of lectures per week and two hours of exercises per week. The studies can be completed by specialized lecture series on data bus systems, EMI and driver assistance systems. Specialized labs on Automotive Network Management and Diagnosis and a summer camp on AUTOSAR tools offer practical training in the field of automotive systems engineering. In the long run the teaching of automotive systems engineering at TU Braunschweig will be intensified in a specialized master program "Electronic Automotive and Aerospace Systems".

References

AUTOSAR: AUTOSAR, Wikipedia, http://en.wikipedia.org/wiki/AUTOSAR, (2012). Accessed 15 Aug 2012

Bergmiller, P.: Design and safety analysis of a drive-by-wire vehicle. In: Maurer, M., Winner, H. (eds.) Automotive Systems Engineering. Springer, Heidelberg (2013)

Bock, T., Maurer, M., Färber, B.: Vehicle in the Loop (VIL)—a new simulator set-up for testing advanced driving assistance systems. In: Driving Simulation Conference North America, University of Iowa (2007)

Bock, T.: Bewertung von Fahrerassistenzsystemen mittels der Vehicle in the Loop-Simulation. In: Winner H., Hakuli S., Wolf G. (eds.) Handbuch Fahrerassistenzsysteme—Grundlagen, Komponenten und Systeme für aktive Sicherheit und Komfort. vol. 1, pp. 76–83. Wiesbaden, Vieweg & Teubner (2009)

Brahmi, M.: Reference systems for environmental perception. In: Maurer, M., Winner, H. (eds.) Automotive Systems Engineering. Springer, Heidelberg (2013)

Breu, A., Holzmann, M., Maurer, M., Hilgers, A.: Prozess zur Komplexitätsbeherrschung bei der Entwicklung eines Stillstandsmanagements für ein hochvernetztes Fahrerassistenzsystem, in Stillstandsmanagement, 8.-9. November 2007, Haus der Technik, Essen (2007)

Darpa: Journal of Field Robotics, Special Issue on DARPA Urban Challenge (Part I–III), Issues 8–10 (2008)

Ehrlenspiel, K.: Integrierte Produktentwicklung: Denkabläufe Methodeneinsatz, Zusammenarbeit. Hanser, München (2007)

Eltaher, A., Maurer, M.: Testing of reconfigurable systems: a cognitive-oriented approach. In: Maurer, M., Winner, H. (eds.) Automotive Systems Engineering. Springer, Heidelberg (2013)

Färber, B., Maurer, M.: Nutzer- und Nutzen-Parameter von Collision Warning und Collision Mitigation Systemen. In: Maurer, M., Stiller, C. (eds.) Workshop Fahrerassistenzsysteme—FAS2005, Walting (2005)

Geyer, S.: Maneuver-based vehicle guidance based on the Conduct-by-Wire principle. In: Maurer, M., Winner, H. (eds.) Automotive Systems Engineering. Springer, Heidelberg (2013)

Glaser, H.: Fahrwerk und Fahrerassistenz - eine ideale Kombination?. In: 7. Symposium Automatisierungs-, Assistenzsysteme und eingebettete Systeme für Transportmittel, AAET 2006, 21–23 Feb, Braunschweig (2006)

Heißing, B., Ersoy, M.: Fahrwerkhandbuch 2nd edn. Vieweg & Teubner Wiesbaden (2008)

ISO: ISO 15622, Transport information and control systems—Adaptive Cruise Control systems—Performance requirements and test procedures (2002)

Kohoutek, P., Dietz, J., Burggraf, B.: Entwicklungsziele und Konzeptauslegung des neuen Audi A4, in ATZ/MTZ extra - Der neue Audi A4, September 2007. Vieweg Wiesbaden (2007)

Kopischke, S.: Entwicklung einer Notbremsfunktion mit Rapid Prototyping Methoden. Dissertation, TU Braunschweig (2000)

Lotz, F.: System architectures for automated vehicle guidance concepts. In: Maurer, M., Winner, H. (eds.) Automotive Systems Engineering. Springer, Heidelberg (2013)

Maurer, M., Behringer, R., Dickmanns, D., Hildebrandt, T., Thomanek, F., Schiehlen, J., Dickmanns, E.D.: VaMoRs-P an advanced platform for visual autonomous road vehicle guidance. In: Proceedings of the Mobile Robots IX (SPIE), pp. 239–248. Boston (1994)

Maurer, M.: Flexible Automatisierung von Straßenfahrzeugen mit Rechnersehen, Fortschritt-Berichte VDI, Reihe 12: Verkehrstechnik/Fahrzeugtechnik, Bd. 443, Ingolstadt (2000a)

Maurer, M.: EMS-Vision: knowledge representation for flexible automation of land vehicles. In: Proceedings of the IEEE Intelligent Vehicles Symposium, Dearborn, 3–5 Oct (2000b)

Maurer, M., Wörsdörfer, K.-F.: Unfallschwereminderung durch Fahrerassistenzsysteme mit maschineller Wahrnehmung - Potentiale und Risiken, Unterlagen zum Seminar Fahrerassistenzsysteme und aktive Sicherheit, Haus der Technik, Essen, 20. November (2002)

Maurer, M.: Entwurf und Test von Fahrerassistenzsystemen im Kraftfahrzeug. Vortrag im Rahmen eines Kolloquiums Realzeit-Computersysteme, TU München, Februar (2007)

Maurer, M.: Forward collision warning and avoidance. In: Handbook of Intelligent Vehicles. Springer, London (2012a)

Maurer, M.: Entwurf und Test von Fahrerassistenzsystemen. In: Winner, H., Hakuli, S., Wolf, G. (eds.) Handbuch Fahrerassistenzsysteme - Grundlagen, Komponenten und Systeme für aktive Sicherheit und Komfort, vol. 2, pp. 43–54. Vieweg & Teubner Wiesbaden (2012b)

Maurer, M.: Vorlesung: Elektronische Fahrzeugsysteme 2, Sommersemester 2012. Braunschweig (2012c)

Ohl, S.: Static software architecture of the sensor data fusion module of the stadtpilot project. In: Maurer, M., Winner, H. (eds.) Automotive Systems Engineering. Springer, Heidelberg (2013)

Reif, K.: Automobilelektronik - Eine Einführung für Ingenieure. ATZ/MTZ-Fachbuch, Vieweg Wiesbaden (2006)

Reschka, A., Böhmer, J.R., Nothdurft, T., Hecker, P., Lichte, B., Maurer, M.: A surveillance and safety system based on performance criteria and functional degradation for an autonomous vehicle. In: Proceedings of the 15th International IEEE Annual Conference on Intelligent Transportation Systems. Anchorage, AK, United States, Sep (2012)

Schäuffele, J., Zurawka, T.: Automotive Software Engineering, 3rd edn. ATZ/MTZ-Fachbuch, Vieweg Wiesbaden (2006)

Wille, J.M., Saust, F., Maurer, M.: Stadtpilot: Driving autonomously on Braunschweig's Inner Ring Road. In: Proceedings of the IEEE International Conference on Intelligent Vehicles (IV). San Diego (2010)

Part II
Requirement Analysis and System Architectures

Chapter 3
System Architectures for Automated Vehicle Guidance Concepts

Felix Lotz

3.1 Introduction and Motivation

Analyzing the current landscape of automotive engineering and the respective research domains leads to the conclusion that vehicle automation is becoming a key technology for the near future. A large part of today's automotive innovation derives from advanced driver assistance systems (ADAS) which are being developed essentially to make driving safer, more comfortable and economically more efficient.

The state of technology counts about 60 available assistance systems for passenger vehicles which help to prevent traffic accidents from happening (Barrios et al. 2007, pp. 9–12). To enable this large number of assistance functionalities, modern cars contain up to 80 electronic control units (ECUs) and a variety of network platforms (Broy et al. 2006, p. V). Furthermore, when looking at an ADAS-Roadmap (Winner and Weitzel 2012, p. 666) it can be assumed that the number of assistance functionalities will increase even more in the future, and will most likely lead to more and more complex systems. The situation is becoming increasingly complicated because of the rising number of manufacturer vehicle models, platforms and powertrain concepts, including different engines and their degree of electrification, in order to fill market gaps and satisfy customer demands for individualization and individual mobilization.

Faced with this variety and complexity in automotive systems design, the engineer and system architect have to deal with challenging problems which cannot be solved through the linear addition of functionalities and control units into already existing architectures, and this not only because of hardware packaging problems (Reichart and Bielefeld 2012, p. 84). The functional variation and diversity mean that the time and effort of system application as well as the costs of corresponding testing and validation will probably increase and can lead to an uneconomical development

F. Lotz (✉)
Institute of Automotive Engineering, Technische Universität Darmstadt, Petersenstraße 30, 64287 Darmstadt, Germany
e-mail: lotz@fzd.tu-darmstadt.de

M. Maurer and H. Winner (eds.), *Automotive Systems Engineering*,
DOI: 10.1007/978-3-642-36455-6_3, © Springer-Verlag Berlin Heidelberg 2013

process. Another aspect is the risk of the visual and mental overload of the driver, who has to interact with and know the system boundaries of the individual assistance systems (Kauer et al. 2010, p. 1214).

From a functional viewpoint, a higher grade of vehicle automation opens the opportunity to incorporate existing systems into an integrated, functionally combined assistance approach and hence provides a possible solution to the problems described above. Presently, many research projects initiated by industry, academia and also the military concern the development of semi- and fully automated driving (cf. Sect. 3.3.1). The latter, often also referred to as autonomous driving, has to address extensive technical and social requirements. Besides the challenge of machine-based perceiving and understanding of a highly complex traffic environment like inner-city driving, to date there is still no valid metric to give a proof of safety, which is required for legal registration and in order to resolve the issue of manufacturer liability (Winner and Weitzel 2012, p. 661). A recent report by the German Federal Highway Research Institute (BASt) comes to the conclusion that highly and fully automated driving, in which the driver has not to control the automated system permanently, is inconsistent with today's German regulatory law because the driver is obligated to show permanent caution in road traffic (Gasser et al. 2012, p. 3).

A legally acceptable intermediate solution for automation concepts could be the vehicle guidance paradigm of "cooperative automation". Cooperative automation can be characterized as an intensive, cooperative interaction between the driver and the automated system based on mutual information, recommendation and approval in order to encounter the driving task more effectively than without one another (Löper et al. 2008; Hakuli et al. 2012, p. 641). For example, in the cooperative assistance concept "Conduct-by-Wire", the system relieves the driver of the vehicle stabilization task and yet enables him/her to stay in the control loop by communicating with the car at maneuver level. Therefore, the driver is still holding responsibility as demanded by law (Hakuli et al. 2011, p. 221; Geyer 2013).

Besides the approach of combining assistance functionalities by an integrated vehicle automation concept, a very important tool to manage the overall system complexity is the system architecture. Not only does it bridge the gap between requirements analysis and implementation by defining the structural layout of the automotive system, but it also accounts for an efficient development process allowing risks to be identified at an early stage of development, enabling the division of labor within a project group and promoting a mutual understanding between all stakeholders (Posch et al. 2007, pp. 14–15).

However, a well-designed system architecture does not by itself generate a successful overall development process. Similar to the functional integration already described above, the upcoming challenges in automotive engineering also require an integrated development process which addresses requirements engineering, the design of test and validation strategies and also tool development. Independent of specific applications, in this chapter such an integrated development process is referred to as 'automotive systems engineering'.

In the following sections we will specify the role of the system, and particularly the software architecture within an integrated development process based on the interdisciplinary research project PRORETA 3. This is the latest project within the PRORETA long-term research initiative between the TU Darmstadt and Continental AG. The objective is to design a virtual safety corridor that prevents accidents without limiting the application to specific use cases. Furthermore, a semi-automated and maneuver-based vehicle guidance concept is being developed in order to relieve the driver from the task of vehicle stabilization. Hereby, emphasis is set on an intuitive close-to-production HMI-solution[1] and the "Safety Corridor" as a solution to combine exclusively intervening and semi-automated assistance systems. In contrast to other research projects, a concept is derived on how this collision mitigation function can be efficiently integrated in synergy into a semi-automated driving concept (Bauer et al. 2012).

One outcome of the project is to implement the PRORETA 3 assistance concept in a test vehicle in order to verify the system performance. The purpose is on the one hand to analyze the warning and intervention strategy within the "Safety Corridor" mode which aims to prevent accidents and critical situations in a 360-degree field of view, and on the other to investigate the vehicle behavior with respect to the maneuver-based, cooperative driving mode, including driver interaction. In this mode, the driver is given the opportunity to assign a vehicle maneuver, like turning left or right at an intersection or going through a roundabout, and then supervise the assistance system that automatically accomplishes the selected maneuver.

In order to achieve the functionalities described above, the test vehicle is equipped with environment sensors as shown in Fig. 3.1. Besides a GPS-receiver, attached radar sensors and their opening angle are indicated in light and dark blue, the stereo camera is indicated in green. Furthermore, the vehicle comes with an active force-feedback gas pedal, controllable electric power steering and a controllable brake booster.

The sensor- and actuator-configuration is an important part of the hardware architecture. Nevertheless, in the following section, focus is set on the software architecture since it offers greater degrees of freedom within the corresponding design process.

An overview over the state of technology for automated vehicle concepts and the respective architectures is given in Sect. 3.3.

3.2 Software Architecture Design

3.2.1 Definitions

According to Vogel et al. (2005, p. 46), there are many definitions for the term "Software Architecture". However, a common definition is the following (Bass et al. 2003, p. 3):

---

[1] **HMI**: Human-Machine Interface.

"The software architecture of a program or computing system is the structure or structures of the system, which comprise software elements, the externally visible properties of those elements, and the relationships among them."

In analogy to the architecture of a building or a house, the architecture is usually displayed in form of "views", comparable to a construction plan. Similar to a building, different views of the architecture can be obtained, for example a blueprint, a plan of electricity or a plan of statics. In software architecture, the four most common (Darms 2007, p. 3) views (or viewpoints) are called "context view", "runtime view", "deployment view" and "module view" (IEEE 2000), while every view is important for a different group of stakeholders (Starke 2009, p. 15).

For the architecture design process, the most important view (and the one most published in literature) is the "module view", which represents the static structure in terms of modules, subsystems, components and the interfaces among them. It is possible to represent the module view in different levels of abstraction, whereas the most abstract view would be the context view (which, for example, shows the interaction of a user with the software) or the most detailed view, which would be the software source code itself (Starke 2009, p. 79).

In the definition of software architecture and in most publications concerning architecture design, the term "system" is often used, yet mostly without a definition. However, in order to understand the difference between the architecture and the system itself, the term "system" has to be defined. In system theory, the following definition can be found (Vogel et al. 2005, p. 52):

"A system is an entity which maintains its existence through the mutual interaction of its parts."

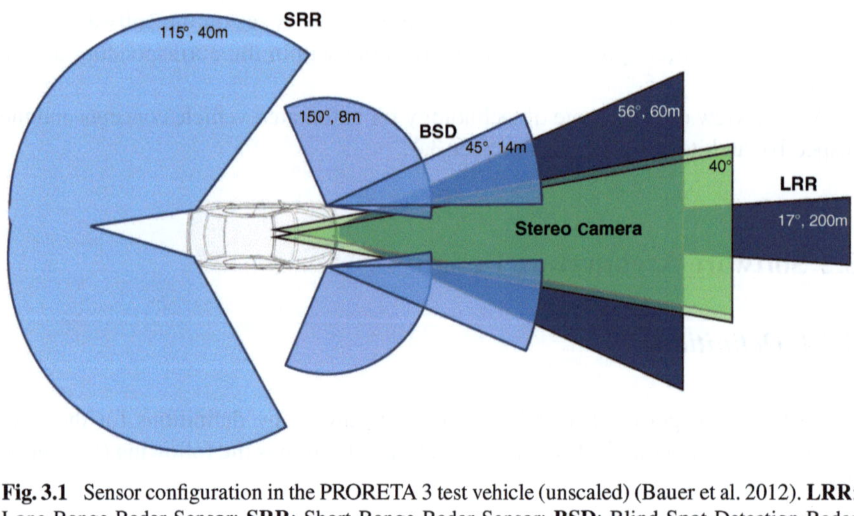

Fig. 3.1 Sensor configuration in the PRORETA 3 test vehicle (unscaled) (Bauer et al. 2012). **LRR**: Lone Range Radar Sensor; **SRR**: Short Range Radar Sensor; **BSD**: Blind Spot Detection Radar Sensor.

According to this definition, a system consists of interacting parts (or modules) and has a system border. It can also be a module for a superior system. Systems exist in order to achieve an objective and therefore exchange information with their environment (open systems) or at least stand in an energetic relation with it (closed systems) (Vogel et al. 2005, p. 53). Hence, the system architecture cannot be seen as the system itself, but as a description of the structure of the system.

Another important finding of system theory is the so-called "emergence", which means that the system possesses characteristics that are more than just the sum of its modules' characteristics. These "emergent" characteristics come into existence by module interaction (Vogel et al. 2005, p. 54). Since the architecture only describes the structure and interfaces between them, by itself it can neither describe the "holistic" system characteristics, nor verify the overall system requirements.

A driver assistance system can be associated with the definitions mentioned above. Its objective is to assist the driver of a vehicle and to this end, exchanges information with its environment, e.g. with HMIs or environmental sensors. While the assistance system consists of interacting software and hardware-modules, their structure and interfaces are described by the system architecture. The emergent characteristics of the overall assistance system have, therefore, to meet the product requirements and are an outcome of the module interaction. Concerning the test- and validation strategy, this means that it is not sufficient simply to evaluate the correct module characteristics and functionalities (which is called reductionism) but also the overall, holistic system characteristics have to be included.

3.2.2 The Role of the Architecture

The significance of the architecture design within the overall development process of a driver assistance system can best be described by the well-known V-Model as shown in Fig. 3.2. The V-Model has proven itself for the top-down and structured development of complex technical systems.

On the left-hand side, the specification branch is depicted, which ranges from the overall product requirements down to detailed software components. The specification and design of the logical, technical and software architecture take place within these development milestones. Hence, as already mentioned in Sect. 3.1, the architecture design is the connecting process between requirements definition and software implementation. A characteristic of the V-Model is that for every specification step a suitable testing strategy is defined, which forms the basis for the component and system validation branch on the right hand side (Fig. 3.2).

The meaning of the architecture for the development team involved has already been described in Sect. 3.1. For the system architect and the software programmer it has another important function: It serves as a "skeleton system" within the implementation phase (Vogel et al. 2005, p. 284). This means that first of all, "empty" software modules and their interfaces are implemented while the module functionalities are extended in later steps. This approach makes it possible to first implement

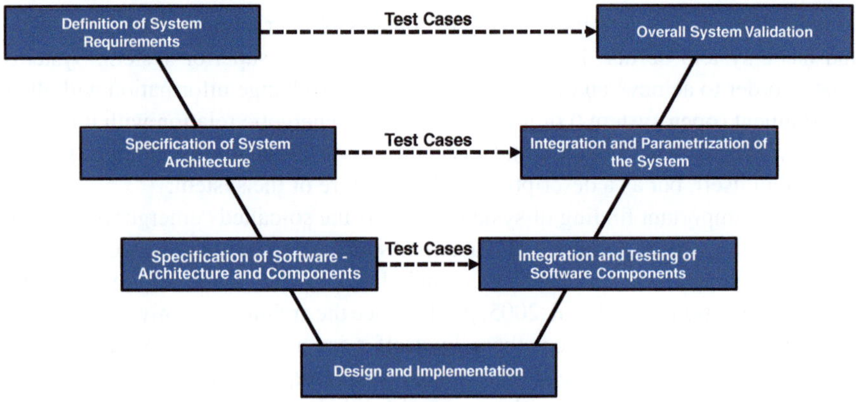

Fig. 3.2 Basic structure of the V-Model, see Maurer (2012, p. 45)

the full functionality of one specific use case (the so-called "cut") and then add the functions needed for further use cases gradually, which allows risks to be identified and simultaneously obtain a functioning system in an early development phase.

Having explained the importance of the architecture design within the development process, the following section will now focus on the procedure of architecture design.

3.2.3 The Architecture Design Process

Figure 3.3 shows the architecture design process, derived from Starke (2009, p. 33) and Posch et al. (2007, p. 57). The first step is the system requirements analysis which is a fundamental process not only for architecture design, but also for the whole development process (cf. V-Model in Fig. 3.2). Requirements can be divided into functional requirements, e.g. specific use cases for the assistance system, and non-functional requirements, e.g. safety-related requirements like a backup strategy during a sensor breakdown, the real-time capability or testability of the system.

In the context of the requirements analysis for driver assistance systems, a scenario-based, use-case-driven approach is proposed as successfully used within the PRORETA 3 project (step 1). The use cases are derived based on the assistance target of the system.

As an example for the cooperative automation mode (cf. Sect. 3.1) use cases are structured by means of basic driving maneuvers for vehicle automation (Tölle 1996). Also, scenario features are specified such as road geometry and the description of the behavior of other vehicles. By describing the desired system behavior within all use case scenarios, a detailed list of requirements can be derived.[2] Other advantages of such a scenario-based use-case description are the clear communication and discus-

[2] See also Vogel et al. (2005, p. 278).

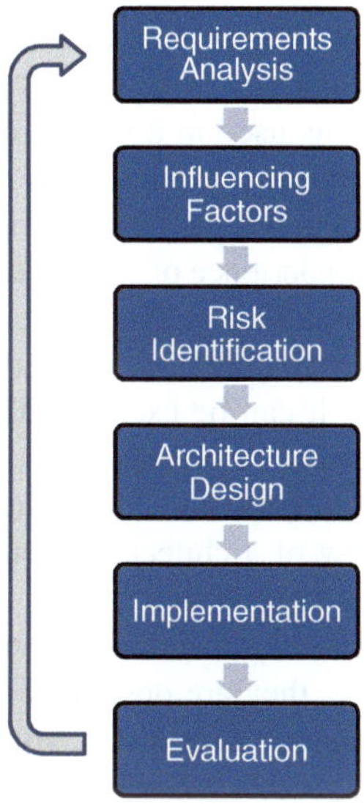

Fig. 3.3 Proposed architecture design process

sion of the system requirements within the project team, the possibility to analyze the scenarios in order to identify the specific system modules and functionalities (e.g. a traffic light detection module) and to use it as a foundation for the later testing strategy ("test cases").

Step two of the architecture design process is to identify the factors influencing the architecture. Organizational factors are, for example, the project budget, time schedule or given programming tools and frameworks (Posch et al. 2007, p. 75). Also technical influencing factors like given hardware components have a big impact on the later system architecture. Limiting factors often come from political issues and policies within the project group or project clients, not necessarily from technical limitations (Starke 2009, p. 48). As an example, an OEM supplier is more eager to set up an ADAS testing vehicle using its own brand of sensors instead of using other sensor concepts from a competitor, even if such concepts would better suit the system in development.

Further along the architecture design process (step 3), potential risks are identified by comparing the system requirements and the limiting influencing factors. As a result, strategies then have to be derived to minimize the impact of the influencing factors. Example negative influencing factors could be the limited availability of important resources like suitable development and testing tools or a strategy to set up an interface to external systems that is insensitive to software versions.

In step four, the actual software architecture is derived. For this, a literature research is necessary in order to identify the state-of-the-art of architecture design for the system in development and to pinpoint similar architecture solutions, architecture patterns or reference architectures. Section 3.3.1 describes the outcome of the research regarding vehicle automation architectures.

Architectural heuristics play an important role within the design process, since they are tools to reduce the complexity of systems with wide degrees of conceptual freedom. The term "heuristics" is used in a variety of meanings throughout architectural literature, e.g. in synonym with the terms "rule", "pattern" or "principle" (Starke 2009, p. 136). In this chapter, the term heuristics is understood as a "code of practice" obtained from the experience of system architects in the past. One common and important heuristic is modularity. The principle of modularity indicates an architecture in which the single module has a distinct and closed functionality and well-defined interfaces, so that it can be exchanged and reused arbitrarily. Modularity is of such importance because it incorporates the also imminent principles of "abstraction", "separation of concerns" and "information hiding" (Vogel et al. 2005, pp. 129–130). A good overview of architectural heuristics can be found in Vogel et al. (2005, p. 111ff.) and Starke (2009, p. 135ff).

Architectural patterns, in contrast, are defined as solutions for recurring problems in software design. In literature, they are described as a practical approach with a distinct solution, often documented in form of UML[3]-diagrams or even code fragments (Vogel et al. 2005, p. 175) and hence are on a more concrete level compared to heuristics. Example patterns are "adapters", "proxies" or "observers" (Starke 2009, p. 167ff).

Given the many examples of architecture heuristics and architecture patterns, it becomes obvious that a thorough literature research on similar systems can pay off quickly in terms of project resources. After the communication and implementation process step of architecture design (cf. the "skeleton system" in Sect. 3.2.2), the architecture design has to be constantly evaluated because requirements as well as influencing factors can change during the development phase, which is therefore characterized as an iterative process (Starke 2009, p. 25).

The evaluation process of architecture design has to answer the question of whether the designed architecture is "good enough". That raises another question: How is a good architecture characterized and is there a suitable criterion or quantity to be measured for its evaluation?

According to Starke (Starke 2009, p. 301), there are two ways to evaluate software projects. The first is to rate organizational aspects of the development process, like the amount of resources needed. However, this kind of evaluation does not provide evidence about the quality of the software product. The second way is to analyze so-called "artifacts" coming from the development process, i.e. requirement lists, architecture views or source code. Only a few of those artifacts are suitable for evaluation using a quantitative criterion, e.g. the number of lines of source code, the amount of memory capacity needed on the computer, the number of test cases needed or, for processes for example, the number of implemented features per time unit.

Therefore, an architecture itself cannot be measured quantitatively. It has to be subject of a qualitative evaluation, that means in respect of its composition and suitability (Starke 2009, p. 302). In detail, the evaluation focuses on how well a designed

[3] **UML**: Unified Modeling Language.

architecture is likely to meet the functional and non-functional requirements derived during the earlier requirement analysis process (Posch et al. 2007, pp. 12–13). Therefore, the requirements in terms of quality criteria are prioritized as to their importance for the assistance system. This results in a "utility tree" (Starke 2009, p. 310). With the help of specific use case analyses, it is then possible to evaluate in a structured way the suitability of the architecture to meet the quality criteria. This "Architecture Tradeoff Analysis Method" (ATAM) has a lot in common with the "house of quality" methodology (Bohn 2012), well-known from product development science.

Although a lot of effort can be put into the evaluation process, it cannot describe the quality of the architecture to the full extent. The missing piece of architectural quality is a phenomenon known in software science under the term "quality without a name" (QWAN). Lacking a precise scientific definition, it can best be described as the unmistakable esthetics, structure and beauty of an architecture (Vogel et al. 2005, p. 176). Although this phenomenon has not been analyzed appropriately, it can be stated that most systems that are not well-structured and are hard to understand tend to be problematic in terms of the fulfillment of their requirements. Hence, the influence of an architecture's esthetics should not be underestimated within the architecture design and evaluation process.

Based on this description of the architectural design process and associated challenges, the following section summarizes the functional and non-functional requirements for the assistance system and sums up important quality criteria to satisfy the requirements for the architecture itself within the PRORETA 3 project.

3.2.4 Requirements for Software Architectures

To be able to conduct a preliminary evaluation of the software architectures identified in the literature research in Sect. 3.3.1 and to structure important quality criteria for the PRORETA 3 software architecture, it is necessary to define a list of the most important architectural requirements. These are:

- Modularity in terms of
  - exchangeability of software modules
  - re-usability of single software modules for other ADAS
  - expendability of system functionalities
  - changeability in terms of robustness against changes and software versions
- Testability in terms of functional testing and debugging that lead to a preferably easy architecture verification and system- or component validation

The requirements above demonstrate that an important goal of the PRORETA 3 project is to develop an ADAS software architecture that is arbitrarily scalable in its functionality in respect to the level of vehicle automation, similar to the work of Darms (2007), who proposed a reference architecture for sensor data fusion for advanced driver assistance systems and Maurer (2000), who introduced a flexible architecture in terms of the degree of automation for vehicles with dynamic machine vision. The issue at hand is how such an architecture has to be composed in order to satisfy the principle of being "as substantial as possible while just as complex as necessary".

3.3 Architectures for Vehicle Automation

The previous section pointed out the need for a literature research to be carried out in order to find out how other ADAS software architectures concerning vehicle automation are structured, which modules are then necessary and how the developers proceeded within other projects.

If the research is limited to the field of driver assistance systems, the software architecture will only play a minor role in most publications.[4] Usually, the architecture itself is depicted as a low detail module view (cf. Sect. 3.2.1) and with regards to content is dwarfed by the systems' functionalities. Unfortunately, little or no explanation is given of how the architecture was derived or what the advantages of the architectural structure are. This stands in a noticeable contrast to publications concerning robotics, where architectural structures and reference architectures seem to be a keen discussion topic.

However, advanced driver assistance systems in terms of "intelligent vehicles" are not in contrast to the robotics discipline but can rather be seen as an application of robotics. This is why the outline given below is structured similar to Kortenkamp and Simmons (2008, p. 187ff), who give a good overview over robotic architectures.

3.3.1 State of Technology

3.3.1.1 Sense-Plan-Act

The sense-plan-act (SPA) paradigm was one of the first architectures applied in robotics. It was employed in mobile robots that used sensors (e.g. cameras) to perceive and build an internal model of their environment, used a planner module to generate a plan in form of a sequence of actions and then executed this plan with the help of its actuators (Kortenkamp and Simmons 2008, p. 189). This linear sequence of events has the disadvantage that the robot is not able to react to unforeseen occurrences like an oncoming obstacle, since it does not use its sensors while moving. Obviously, therefore, this sequential architecture is not suitable for a dynamic environment like road traffic.

[4] This finding is also described in Maurer (Maurer 2000, p. 15).

3.3.1.2 Reactive Architectures

Reactive Architectures emerged after the sense-plan-act paradigm reached its limits; they are characterized by quick, "reactive planning". Reactive architectures exist of different interacting finite state machines which are called "behaviors" (Kortenkamp and Simmons 2008, p. 190). This is why this kind of architecture has embossed the term "behavioral robotics". Behaviors access sensors and actuators during their runtime so they can react to changes in their environment quickly, just like a "reflex". This is also why robots based on this architecture are said to have an "insect-like" behavior. An example of a mobile robot is a small robotic vacuum cleaner with the behaviors "wander around" and "collide". Every time the robot collides with a wall, the "wander" behavior is subsumed by the "collide" behavior and the heading angle is changed randomly. After that, the "wander" behavior is active again. A well-known type of reactive architectures is the "subsumption architecture" of Brooks (1986). In this architecture, it is possible to just add more and more behaviors in order to create more complex robots.

However, this also brings disadvantages: Adding more behaviors leads to a complex state machine interaction and, since only a few behaviors can be active at the same time, results in an arbitration problem (Kortenkamp and Simmons 2008, p. 190). An easy way to deal with multiple behaviors is to categorize low- and high-level behaviors, like a hierarchy. In the vacuum cleaner example, the "collide" behavior is superior to the "wander around" behavior. Another way to arbitrate different behaviors was proposed by Arkin (1987) with the "motor-schema" based on human perception and action. In it, every behavior creates a potential field as an overlay over the current environment. The robot's path then is derived from the sum of all of those potential fields. In order to also fulfill more complex tasks, a high-level planning module was implemented that consists of a mission planner, a navigator that calculates waypoints and a pilot that controls the different behaviors.

A further way to deal with the arbitration problem was proposed by Rosenblatt (1997) with the "Distributed Architecture for Mobile Navigation" (DAMN). It also consists of different behaviors, however, an arbitration module on top assigns different priorities to them, depending on the current situation. The behaviors then each vote for a path curvature they want the vehicle to execute in the future while being weighted with the corresponding priorities. Instead of using a mean curvature of all votes, the arbiter, in contrast to Arkins' scheme, chooses the path with the most votes (cf. Fig. 3.4). This way, inherent problems of averaging commands, that can lead to unsatisfactory results like logical minima on potential fields, are avoided (Rosenblatt 1997, p. 343).

Another advantage of this architecture design is the directly resulting vehicle trajectory, since the curvature and therefore the maximum safe vehicle velocity is a direct outcome of the arbitration process. As an example, a successful implementation of this architecture approach was conducted by the German team "Caroline" in the DARPA urban challenge. The autonomous vehicle concept used the DAMN arbitration concept and combined it with an interrupt function to be able to also react to road intersections, roadblocks and other events (Rauskolb et al. 2008, p. 701).

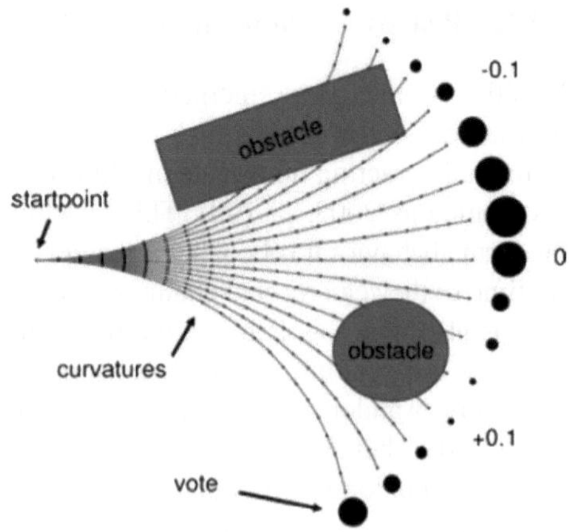

Fig. 3.4 Result of the DAMN arbitration process (Rauskolb et al. 2008, p. 701)

To sum up, reactive architectures have the advantage that they are able to react quickly to unforeseen events and can be easily expanded in terms of the system's functionality. However, this architectural style reaches its limits in complex environments like road traffic since the interaction of the behaviors becomes hard to arbitrate and to optimize. Also, no long-term planning is available such as the planning to drive a complex route. The examples of the Arkin scheme as well as the DAMN architecture show that there has to be a superior module to coordinate the individual behaviors. This circumstance leads to the so-called "layered architectures" explained below.

3.3.1.3 Layered Architectures

Layered architectures emerged in order to combine the advantage of reactivity with deliberate planning. The most common type of layered architectures has three hierarchical layers, as depicted in Fig. 3.5. Just like reactive architectures, in the lowest layer a set of behaviors is implemented, which have the closest connection to sensors and actuators. These behaviors are either active all the time and have to be arbitrated or are triggered specifically by the higher-level executive layer, for example when two behaviors compete for the same resource (e.g. an actuator) (Kortenkamp and Simmons 2008, p. 196). The task of the executive layer is to decompose a high-level plan coming from the planning layer into a sequence of low-level behaviors. Therefore, it has to also set temporal constraints if behaviors are conducted in a sequence or are active concurrently. Usually, the executive layer is implemented as a hierarchical finite state controller which also has the function to monitor the behavior execution and to handle exceptions within a routine (Kortenkamp and Simmons 2008, p. 198).

The intention of the high-level planning layer is to generate long-term plans to reach the system goal. Also, it has to re-plan in case a situation changes, e.g. if a road is blocked. To do this, planners usually can be seen as schedulers, which lay out

Fig. 3.5 Exemplary structure of layered architectures (Kortenkamp and Simmons 2008, p. 191)

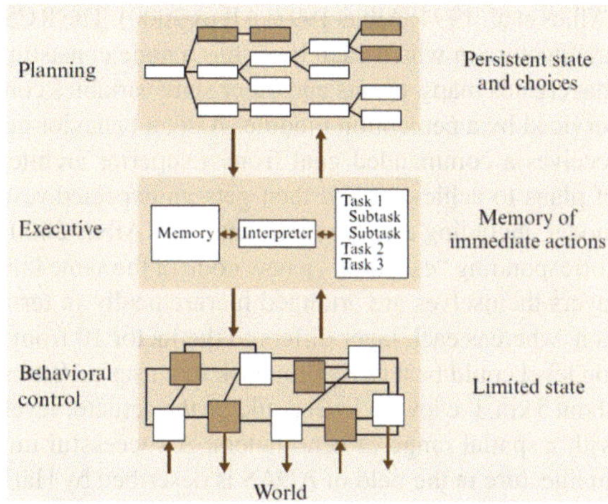

tasks for the executive layer on a time line (Kortenkamp and Simmons 2008, p. 199). They can be implemented in that the executive layer requests a task at a given time or that the planner is always active and hands out tasks to the executive layer.

Maurer (2000, pp. 32–34) summarizes three different types of theoretical hierarchical decompositions. One possibility is to use a "hierarchy of description", in which a complex problem can be described in various levels of detail. In it, a high level layer describes the problem's meaning in order to get a deeper understanding of it and low-level layers give a more detailed explanation. Another type is the "hierarchy of decisional layers" in which the solution to a complex problem can be substituted with a sequence of low order problems that have to be solved. A third way of hierarchical decomposition is the "hierarchy of organization" for multi-goal systems. This kind of architecture is characterized by layers which do not control the lower-order layers completely but coordinate them in such a way that they have a specified freedom of action. The advantage of this type of hierarchy is a more efficient decision process, since the higher-order layers only have to intervene in case a conflict of goals occurs. Hence, such a hierarchy supports the ability to introduce specialist modules or agents into the architecture.

Besides combining reactivity and deliberate planning, layered architectures have the advantage that each layer can be developed, implemented and tested independently of each other. For example, in an early stage of development, behaviors can be triggered manually and tested without the need for an executive layer, or an executive layer state machine can be checked by the trigger outputs in certain situations coming from a planning task without the vehicle or robot needing to actually conduct the behavior.

An important example of a layered architecture is the "Real-Time Control System Architecture" (RCS) which was derived from the US National Institute of Standards and Technology as an advancement of the "NASA Standard Reference Model Architecture" (NASREM) used in various robots, autonomous vehicles and in space travel

(Albus et al. 1994; Albus 1997; Albus 2000). The RCS architecture is a multi-layered architecture in which each layer has a node consisting of a world modeling module that creates maps, events and other state variables concerning its environment and is serviced by a perception module. Also, a behavior generation module exists, which receives a commanded goal from a superior architectural layer, creates a number of plans to achieve it and then gets an expected result of the plan from the world model, including a cost/benefit analysis (Albus 2000, p. 3262). For each plan and a corresponding "executor", a new node of the same kind as just described is used. The layers themselves are arranged hierarchically in terms of spatial range and resolution, whereas each layer differs to the factor 10 from the others. As an example, the top level could be a navigation task that uses a 10 min horizon and a spatial range of about 5 km, the lowest level could be the actuator level that uses a 20 ms time horizon with a spatial range of centimeters. A successful implementation of this reference architecture in the field of ADAS is described by Häring et al. (2009), who proposes the function of an autonomous emergency brake using the RCS paradigm.

A milestone in respect to autonomous vehicle concepts was achieved within the PROMETHEUS project, which introduced the spatio-temporal "4-D" approach (3D space plus time) for dynamic machine vision (Dickmanns et al. 1994). The structure of the hierarchical layers within the 4-D architecture design was inspired by the well-known Rasmussen model of human action (Rasmussen 1983) which comprises "knowledge-based", "rule-based" and "skill-based" behaviors as a base for hierarchical decomposition (Maurer 2000, p. 27). In a subsequent version of the 4-D approach, Maurer (2000) expands the 4-D architecture in a way that various levels of automation could be achieved. Therefore, a key requirement is that the automated system is aware of its own capabilities and is able to activate matching vehicle behaviors during the system's runtime. It works in such a way, that the driver-requested level of automation is permanently compared with correspondent and therefore required quality criteria that have to be fulfilled by the automated system. In case the defined criteria are not fulfilled, for example the dead time of the vehicle control loop exceeds a required threshold, a flexible behavior decision module, which consists of hierarchical ordered state machines, is able to activate appropriate automated driving functions for a lower level of automation. The capability of the 4-D concept was shown by the implementation into the test vehicle "VaMP" which travelled more than 1,600 km in an automated driving mode on public roads (Maurer 2000, p. 114).

The success of both approaches described above (RCS and 4-D) led to a combined architecture which shows the compatibility of both designs, the "4-D/RCS architecture" (Albus 2000). Maurer (Maurer 2000, p. 38) however concludes that a combination of both paradigms results in conceptual disadvantages and hence they should rather be used in parallel, depending on the task to be achieved.

Most recent projects regarding vehicle automation are, for example, the "autonomous car project" of Google or the German team "AutoNOMOS Labs" with the prototype vehicle "MadeInGermany", which has gathered a lot of media attention. Unfortunately, only few details are known so far in terms of the software architecture design of these projects. At least for the "MadeInGermany" project it can be stated that also a layered architecture design was chosen that consists of two main layers, the "behavior layer", itself divided into a high-level "strategy module" which processes data from digital maps and a lower-level "tactics" module which calculates an appropriate vehicle trajectory, and the "controller layer" which controls the given trajectory.[5]

Other examples of layered architectures are widespread and very similar. For further reference see Baker et al. (2008) who used a three-layered architecture for the winning autonomous vehicle "BOSS" in the DAPRA Urban Challenge or Payton (1986), Simmons (1994), Laugier et al. (1998), Miura et al. (1997) and Nelson (1999).

All of the mentioned architectures have been designed for autonomous or highly-automated vehicle guidance. However, since the PRORETA 3 project's intention is to design a vehicle for semi-automated, maneuver-based vehicle guidance, a lacking feature of all of these architectures is that a driver interaction is not provided or not sufficiently described. The state of technology so far regarding automated vehicle guidance in combination with a systematic driver interaction concept consists of the research projects "Conduct by Wire" (CbW), "H-Mode" and "HAVEit".

The software architecture of CbW has a lot in common with the layered architectures described above (Geyer 2013). It is hierarchically structured according to the three-level hierarchy of the driving task of Donges (2012, p. 16), who divides the driving task into the navigation level, the trajectory-based guidance level and the vehicle stabilization level. The architecture consists of a maneuver management module that assigns and enables a set of driving functions, which can be seen as vehicle behaviors, in accordance with the driver's desired maneuver. These maneuvers are obtained via an HMI called maneuver interface. To sum up, in CbW the driver is able to interact with the automation concept via an HMI that allows him/her to assign maneuvers, parameterize trajectories and also manually control the vehicle in an unstructured environment like a parking lot. These interactions then influence the automation in respect to the activation of vehicle behaviors.

In contrast to CbW, the research projects "H-Mode" and "HAVEit" propose a variable level of vehicle automation, ranging from manual driving through to autonomous driving (Löper et al. 2008). The architecture itself consists of a driver interface module, which assesses the driver's state, a command layer to define maneuvers and vehicle trajectories dependent on the automation level and the execution layer which controls the vehicle actuators (Zeng 2010, p. 1667). The driver interaction is arbitrated via a selection unit, in which a quantitative measure, the "valential", decides over the future trajectory. The magnitude of this unit is influenced by a negotiation process between driver and automation based on haptic feedback, e.g. the driver's torque of the steering wheel (Löper et al. 2008, p. 16).

A further publication on driver-vehicle interaction was proposed by Bayouth et al. (1997), who introduced the idea of different automation levels and maneuver based driving already in the year 1997.

[5] Lecture slides and personal question to Professor Raùl Rojas at his speech at FZD, TU Darmstadt on 9 July 2012.

3.3.2 Summary and Discussion

The above excerpt of the state of technology on software architectures of vehicle automation concepts aims to find out which architectural patterns have proved themselves within the field of ADAS and robotics and which system modules are therefore necessary.

It can be stated that the paradigm of behavior-based, layered architectures have proved themselves to be a beneficial architectural pattern for vehicle automation concepts since they have been successfully used in key research projects within the last two decades, right up to the most recent prototypes like "MadeInGermany". Important advantages are the inherent structured arrangement based on a hierarchical and modular task composition, the ability to develop and test the different architectural layers independently of each other and the fact that this type of architecture is capable of deliberate, long-term planning.

An interesting finding is that most layered architectures for vehicle automation consist of three architectural main layers. As Maurer (2000, p. 34) summarizes, three layers have become established in this field of robotics although there is not necessarily a reason for that from the viewpoint of functional architecture. However, we assume that this number comes from common agreed-upon models that help the architect to decompose the driving task in a structured way, like the Rasmussen model of human activities or the Donges model for the driving task, which each contain three main layers (Rasmussen 1983; Donges 2012).

Important modules, in addition to a sensory processing module, that can be found in most automation concepts are a world model that gathers information about the systems' environment, a high-level planning module to be able to achieve long-term goals like vehicle navigation, an executive module that coordinates low-level vehicle behaviors and reacts to exceptions, a set of behaviors and—for cooperative automation concepts—a suitable HMI.

The three projects on cooperative vehicle automation described in Sect. 3.3.1.3 have in common, that no direct mechanical connection exists between the driver and the vehicle because "by-wire" interfaces are being used for the steering as well as the brake and gas pedals. Within the PRORETA 3 project, however, the test vehicle is equipped with regular controllable power steering as well as a brake pedal connected to the main braking cylinder by the electro-pneumatically actuated booster. This leads to the question of how the driver interaction influences the software architecture of automated vehicle guidance concepts when the driver input comes from the hardware architecture.

As a conclusion drawn from the literature research, it can be stated—in accordance with Dickmanns (2005, p. 224) and Gasser et al. (2012, p. 25)—that there is a significant demand for research into the interaction of the human being with automated vehicle concepts. Especially the driver interaction as a problem of command arbitration and its impact on the software architecture and dependency on the hardware architecture is to be investigated.

Based on the information given in this section, an initial design of the PRORETA 3 software architecture is derived in Sect. 3.4.

3.4 Software Architecture for Vehicle Automation

Based on the findings from literature research, the functional requirements of PRORETA 3 and the requirements for the software architecture itself, an architecture design for automated vehicle concepts is derived. In contrast to the robotic architecture patterns for autonomous vehicles, focus is also placed on the human-machine interaction.

A layered architecture is believed to be most appropriate as a design basis for PRORETA 3, in line with the "modularity" and "testability" requirements (cf. Sect. 3.2.4), the inherently well-structured and functionally scalable approach as well as the ability to cope with complex situations such as road traffic. Based on this decision, the question has to be answered of how many layers are, therefore, necessary and how the complex problem of vehicle automation can be decomposed into them.

As shown in Maurer (2000), the model of human activity according to Rasmussen (1983) has proven to be well-suited for an example decomposition, especially for behavior-based layered architectures. This comes from the fact that in behavioral robotics, behaviors can be seen as basic state machines containing mostly simple transfer functions that describe the robot's desired movement and have direct access to sensors and actuators (cf. Sect. 3.3.1.2). Accordingly, Rasmussen (1983) describes the "skill-based behavior" level of human action as a set of automated sensorimotor patterns that process sensory input features into direct actions, which is a very similar procedure. The next higher layer in Rasmussen's model is described as the "rule-based behavior", in which the human recognizes specific "signs" from the perceived sensory features, associates an appropriate task and chooses between a set of stored rules to solve this task. This can be seen as a feedforward control for the sensorimotor patterns (Maurer 2000, p. 29). As an equivalent, most architecture designs use an "executive" or "coordination" layer in which a superior module controls the behaviors in a way that they trigger specific behaviors to solve a recognized task. The similarity of Rasmussen's model to layered architectures also applies for the top-layer. In both paradigms, a planning entity decides between tasks in order to reach its overall goal concerning specific boundary conditions. To sum up, both paradigms consist of hierarchically structured layers that can be understood as layered control loops with an increasing time constant, in which higher layers provide a feedforward control to the next lower layers. Hence, the Rasmussen model serves as a good basis for hierarchical decomposition and is considered in the following design.

After having defined the basic architecture layout and an approach how to decompose it into specific layers, the modules within the layers have to be determined. Tasks that are necessary for vehicle automation are, independent of their exact specification or name, the perception of the vehicle's environment, and an interpretation of the perception that serves as a basis for a behavioral decision which itself feeds a behavior generation (or action) module. A task neglected in most publications on architecture is a description of how to embed the driver into a vehicle automation concept. Therefore, a suitable HMI concept is needed and should be considered in the design.

Based on the above considerations, the actual architecture design process results in a top module view that describes the static architecture layout as shown in Fig. 3.6.

As already described, the architecture is based on three horizontal main layers within the application range (Bauer et al. 2012), which are structured according to their level of abstraction in the Rasmussen model. Vertically, there are also three columns. A world model column has the task of providing all necessary information coming from the vehicle's environment to the planning modules in order to make situation-appropriate decisions. The world model itself is fed by a solution-independent **sensorics- and sensordata-fusion** module which provides information about the existence of perceived objects and the vehicle's surroundings in a detailed but mostly not situational-interpreted way. Approaches of Darms (2007) are suitable for an implementation, however, in PRORETA 3 a geometric grid-map-based description of the static environment as well as an object list for the dynamic environment was chosen. According to the Rasmussen model, most information is used by the lowest "behavior" layer, but also more abstract information via digital maps or Car2X can be provided that a human would not be able to perceive him/herself. From this, the sensorics- and sensordata-fusion module spreads over all layers. On

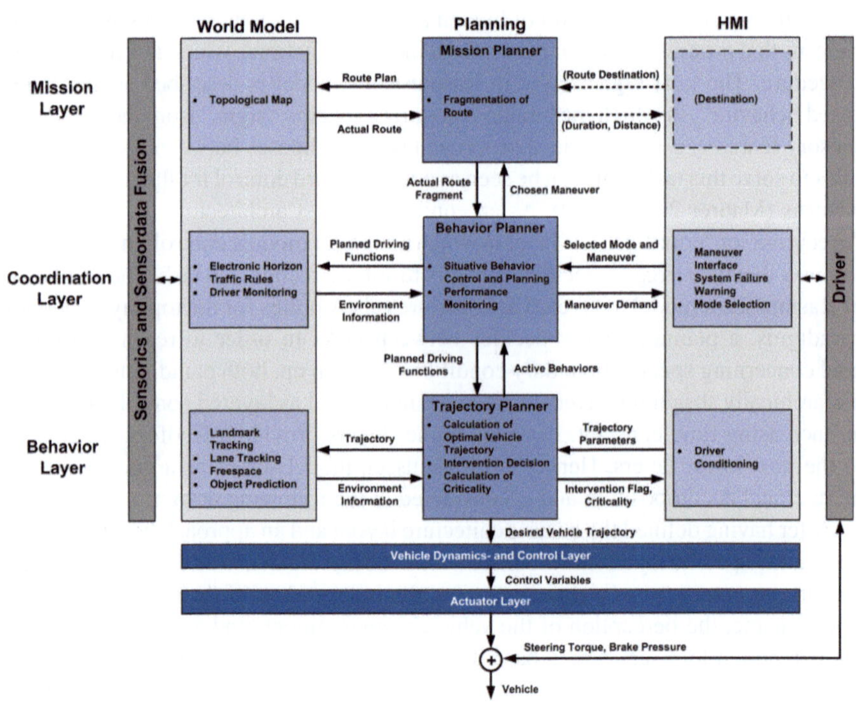

Fig. 3.6 Proposed software architecture for vehicle automation concepts (cf. (Bauer et al. 2012))

this basis, the **world model** is able to interpret the given data in order to analyze and understand a traffic situation. Therefore, specialist modules are needed. In the behavior layer, a landmark tracking module detects localization-relevant features like road junctions or stop lines, a lane tracking module interprets video data for a lane model, a drivable freespace module determines valid maneuvering space (Schreier and Willert 2012) and an object prediction module calculates future positions of relevant entities. For the "coordination layer" more abstract information is needed. Corresponding modules are an electronic horizon based on geometric and digital maps with free-configurable attributes, a module to determine the traffic rules such as the right of way on intersections and the driver monitoring to check whether the driver is able to stay in the overall control loop. For the "mission layer", a graph-based topological map is proposed.

The **planning modules,** which are explained below in this section, receive high-order tasks and split them up into lower-order tasks. A novel element within the proposed behavior-based, layered architecture is the **HMI** column that is needed for driver interaction within a cooperative automation concept. Specialist modules are the driver conditioning which helps to support and warn the driver in critical traffic situations and, for the coordination layer, a maneuver interface to offer the driver possible situation-dependent vehicle maneuvers as well as the possibility to choose a specific mode of automation. A possible module on mission layer level could be the input of a destination, which, however, is not implemented within PRORETA 3.

In the following, the layers and their functionality are described. The layer with the highest hierarchy is called **"mission layer"**. In it, the mission planner module splits the current vehicle route into route fragments and delegates a matching task to the secondary behavior planner module, e.g. "handle intersection" or "follow road". The mission planner module receives its information from a topological map, in which intersections are represented as knots and roads as conjunctive edges. Additionally, corresponding characteristics, e.g. the number of outgoing lanes at an intersection, are deposited. In line with the Rasmussen model, the mission planning module should be able to detect and react to unknown situations in which no rules for the behavior planning module are available. However, in most architecture designs, the task of the top-level layer is limited to the navigation task since—in contrast to human behavior—it is very easy to implement a large number of rules for behavior planning (Maurer 2000, p. 46). For autonomous vehicle concepts, a vehicle destination comes from the HMI module, however, this is not necessary for the cooperative PRORETA 3 concept (dashed lines in Fig. 3.6).

The second layer is called **"coordination layer"** and has the job of controlling the underlying vehicle behaviors according to its current task. The main module within this layer is the behavior planner, which triggers appropriate vehicle behaviors in a timely manner and hence is the central behavior decision unit within the architecture. A common implementation for this kind of planning module is one or more state machines. As an interface towards the behavior layer, disjunctive longitudinal and lateral driving functions, e.g. "lane change" or "follow road" are a possible solution (Maurer 2000, p. 52; Hakuli et al. 2010). As an example of the behavior planner module, the task "handle intersection" with the attribute "traffic light" may result in reducing the vehicles' speed in advance, requesting a desired maneuver from the driver and stopping the vehicle in case of a red traffic light. Therefore, it needs information from the world model, like the right of way or the geometric relation between the car and the scene, e.g. the number of lanes on a road or the distance to the next intersection. Another major task of the behavior planning module is to monitor the system's performance in terms of important resources and modules that are needed for a given task. In case one of those fails or necessary information cannot be provided, it has to transfer the system into a safe state.

The third layer is called **"behavior layer"**, and is where the different vehicle behaviors are implemented. The main trajectory planner module has the function of processing a safe vehicle trajectory in respect to the given driving functions that have been triggered. Therefore, it is possible to implement the behaviors independently, e.g. a sigmoid function for a lane change or a braking intervention in terms of a high negative acceleration, or, to have more complex behaviors like the safety corridor function, that cannot be implemented as just a transfer function. In the trajectory planning module, an emergency trajectory is determined with the use of a potential-field based approach. In order to perform this task, the world model has to provide much more specific information in a spatial and timely manner coming from the specialist modules described above. The HMI on the behavior level is called "driver conditioning" and has the task of embedding the driver not only via information and warning but also in an accident-preventive way.

In the lower hierarchy of the proposed architecture two more layers exist outside of the functional range. In the **vehicle dynamics and control layer**, the given vehicle trajectory is processed into the vehicle dynamics controllers. This layer then generates commands for the actuators like a steering torque or a brake pressure which are then processed by the **actuators** themselves.

The importance of these two layers should not be underestimated since the driver is able to directly influence the actuators depending on the hardware concept of the car. This may not be problematic within an autonomous concept using by-wire interfaces for human-machine interaction or also for single ADAS functionalities like adaptive cruise control. However, it is problematic within the full range of automation functionalities in-between, especially for systems that utilize standard HMIs like a steering wheel and a brake pedal. As an example, in the case of an emergency maneuver, the driver wants to evade to the right, the automation to the left. The question then is, is the driver treated as a disturbance quantity which the automation wants to compensate? If not, how is the force-transition process back to the driver to be defined? So far, no scientific investigation has been conducted on how to integrate the drivers' intention in that direct mechanical access to the actuators has to be integrated within the overall arbitration strategy. This reveals a clear demand for research in the future.

In the architecture design for vehicle automation concepts proposed above, the Rasmussen (1983) model for human action was used as a basis for hierarchical decomposition. A novel element within behavior-based layered architectures is the integration and description of a suitable driver interface into the design, which is especially important for the cooperative automation concepts explained in Sect. 3.1.

Details of the necessary specialist modules for situation interpretation were presented. The proposed design makes some exceptions with regard to the Rasmussen model, e.g. the sensory input and the knowledge-based behavior. However, this is because vehicle behavior cannot be completely subsumed by human behavior (Maurer 2000, pp. 30–31). Decomposition paradigms need to serve as a flexible helper, not as a rigid burden for the system architect.

3.5 Summary and Outlook

This chapter proposes a systematic and top-down development methodology for automated vehicle guidance concepts and their corresponding system- and software architecture.

To this end, a software architecture design process was introduced in Sect. 3.2, which bridges the gap between requirements analysis and implementation within the overall development process according to the well-known V-Model. Explanations were given on the importance of knowing limiting influencing factors, ways of evaluating architecture design and on the conduct of a literature research on architectural patterns.

The state-of-technology identified was that most vehicle automation concepts utilize behavior-based, hierarchically layered architectures since they are well-structured in an inherent way and hence facilitate a modular design. This allows a division of labor and an efficient design-, implementation and validation process. However, most architectural patterns have been designed for autonomous vehicle guidance and do not support a human-machine interaction on a cooperative basis. This is why the architecture proposed in Sect. 3.4 facilitates a new human-machine interface column. From a first evaluation, the architecture meets the functional and architectural requirements mentioned in Sect. 3.2.4 thanks to its modular approach. In future, the proposed architecture has to prove itself within the further development process of the PRORETA 3 project.

A clear demand for research has been identified in terms of the interaction and arbitration metric between the driver and the automation with regard to the system's hardware architecture.

In the automotive systems engineering discipline and the associated scientific community, it would be desirable for a lively discussion to arise about efficient development processes and structures of system architectures, just like that underway in the example discipline of robotics.

Acknowledgments We thank Continental AG for kindly funding this work within the PRORETA 3 cooperation, which aims to develop future concepts for integrated driver assistance systems.

References

Albus, J., Lumia, R., Fiala, J.: NASREM—The NASA/NBS Standard Reference Model for Telerobot Control System Architecture, discussion paper. National Institute of Standards and Technology, Gaithersburg (1994)

Albus, J.: The NIST real time control system (RCS): an approach to intelligent systems research. J. Exp. Theore. Artif. Intell. **9**(2–3), 157–174 (1997)

Albus, J.: 4-D/RCS reference model architecture for unmanned ground vehicles. In: IEEE International Conference on Robotics and Automation. San Francisco, USA (2000)

Arkin, R.: Motor Schema based Navigation for a Mobile Robot. In: IEEE International Conference on Robotics and Automation. Amherst, Massachusetts (1987)

Baker, C., Ferguson, D., Dolan, J.: Robust Mission Execution for Autonomous Urban Driving, Carnegie Mellon Research Showcase,Robotics Institute, Paper 178 (2008)

Barrios, J., Aparicio, A., Dundar, S., Schoinas, D.: Common Database of Existing Safety Functions and Corresponding System Platforms, Report of Deliverable 6.1, Trace Project (2007). http://www.trace-project.org (September 17th, 2012)

Bass, L., Clements, P., Kazman, R.: Software Architecture in Practice, 2nd edn. Addison-Wesley, Boston, USA (2003)

Bauer, E., Lotz, F., Pfromm, M., Schreier, M., Abendroth, B., Cieler, S., Eckert, E., Hohm, A., Lüke, S., Rieth, P., Willert, V., Adamy, J., Bruder, R., Konigorski, U., Winner, H.: PRORETA 3: An Integrated Approach to Collision Avoidance and Vehicle Automation. In: at-Automatisierungstechnik, 12/2012. Oldenbourg Wissenschaftsverlag (2012)

Bayouth, M., Nourbakhsh, I., Thorpe, C.: A hybrid human-computer autonomous vehicle architecture. In: Third ECPD International Conference on Advanced Robotics, Intelligent Automation and Control, Belgrade, Serbia (1997)

Bohn, A.: Produktinnovation, Lecture Notes. Department of Mechanical Engineering, TU Darmstadt (2012)

Brooks, R.: A robust layered control system for a mobile robot. IEEE J. Robot. Autom. **2**(1), 14–23 (1986)

Broy, M., Krüger, I., Meisinger, M.: Automotive Software—Connected Services in Mobile Networks. Springer, Berlin (2006)

Darms, M.: Eine Basis-Systemarchitektur zur Sensordatenfusion von Umfeldsensoren für Fahrerassistenzsysteme. VDI, Düsseldorf (2007)

Dickmanns, E.D., Behringer, R., Dickmanns, D., Hildebrandt, T., Maurer, M., Thomanek, F., Schiehlen, J.: The seeing passenger car 'VaMoRs-P'. In: IEEE Intelligent Vehicles Symposium, Paris (1994)

Dickmanns, E.D.: Vision: Von Assistenz zum Autonomen Fahren. In: Maurer, M., Stiller, C. (eds.) Fahrerassistenzsysteme mit maschineller Wahrnehmung. Springer, Berlin (2005)

Donges, E.: Fahrerverhaltensmodelle. In: Winner, H., Hakuli, S., Wolf, G. (eds.) Handbuch Fahrerassistenzsysteme. 2nd edn. Vieweg/Teubner, Wiesbaden (2012)

Gasser, T., Arzt, C., Ayoubi, M., Bartels, A., Bürkle, L., Eier, J., Flemisch, F., Häcker, D., Hesse, T., Huber, W., Lotz, C., Maurer, M., Schumacher, S.-R., Schwarz, J., Vogt, W.: Rechtsfolgen zunehmender Fahrzeugautomatisierung. Wirtschaftsverlag NW, Bremerhaven (2012)

Geyer, S.: Maneuver-based vehicle guidance based on the Conduct-by-Wire principle. In: Maurer, M., Winner, H. (eds.) Automotive Systems Engineering. Springer-Verlag Berlin, Heidelberg (2013)

Häring, J., Wilhelm, U., Sailer, U.: Systemarchitektur des Predictive Safety Systems, in ATZ Elektronik, 3/2009. Springer, Heidelberg (2009)

Hakuli, S., Kluin, M., Geyer, S., Winner, H.: Development and Validation of Manoeuvre-Based Driver Assistance Functions for Conduct-by-Wire with IPG CarMaker, FISITA 2010 World Automotive Congress. Budapest, Hungary (2010)

Hakuli, S., Geyer, S., Winner, H., Henning, J.: Integriertes Konzept für die manöverbasierte Fahrerassistenz, in ATZ Automobiltechnische Zeitschrift, 3/2011. Springer, Heidelberg (2011)

Hakuli, S., Bruder, R., Flemisch, F., Löper, C., Rausch, H., Schreiber, M., Winner, H.: Kooperative Automation. In: Winner, H., Hakuli, S., Wolf, G. (eds.) Handbuch Fahrerassistenzsysteme. 2nd edn. Vieweg/Teubner, Wiesbaden (2012)

IEEE Standard 1471–2000: Recommended practice for architectural description of software-intensive systems, Institute of Electrical and Electronics Engineers/Circuit Theory Group (2000)

Kauer, M., Schreiber, M., Bruder, R.: How to conduct a car? In: IEEE Intelligent Vehicles Symposium 2010. San Diego (2010)

Laugier, C., Fraichard, T., Paromtchik, I.E., Garnier, P.: Sensor-Based Control Architecture for a Car-Like Vehicle, IEEE International Conference on Intelligent Robots and Systems, Victoria (1998)

Löper, C., Kelsch, J., Flemisch, F.: Kooperative, manöverbasierte Automation und Arbitrierung als Bausteine für hochautomatisiertes Fahren, in Automatisierungs-. Assistenzsysteme und eingebettete Systeme für Transportmittel, Gesamtzentrum für Verkehr, Braunschweig (2008)

Maurer, M.: Flexible Automatisierung von Straßenfahrzeugen mit Rechnersehen, VDI Fortschrittsberichte Reihe 12 Nr. 443, VDI, Düsseldorf (2000)

Maurer, M.: Entwurf und Test von Fahrerassistenzsystemen. In: Winner, H., Hakuli, S., Wolf, G. (eds.) Handbuch Fahrerassistenzsysteme. 2nd edn. Vieweg/Teubner, Wiesbaden (2012)

Miura, J., Ito, M., Shirai, Y.: A three-level control architecture for autonomous vehicle driving in a dynamic and uncertain traffic environment. In: IEEE Conference on Intelligent Transportation System, Boston (1997)

Nelson, M.: A design pattern for autonomous vehicle software control architecture. In: IEEE International Computer Software and Applications Conference, Phoenix (1999)

Payton, D.: An architecture for reflexive autonomous vehicle control. In: IEEE International Conference on Robotics and Automation, San Francisco (1986)

Posch, T., Birken, K., Gerdom, M.: Basiswissen Softwarearchitektur, 2. edn. dpunkt, Heidelberg (2007)

Rasmussen, J.: Skills, rules and knowledge; signals, signs and symbols, and other distinctions in human performance models. IEEE Trans. Syst. Man Cybern. **13**(03), 257–266 (1983)

Rauskolb, F., Berger, K., Lipski, C., Magnor, M., Cornelsen, K., Effertz, J., Form, T., Graefe F., Ohl, S., Schumacher, W., Wille, J.-M., Hecker, P., Nothdurft, T., Doering, M., Homeier, K., Morgenroth, J., Wolf, L., Basarke, C., Berger, C., Gülke, T., Klose, F., Rumpe, B.: Caroline: an autonomously driving vehicle for urban environments. J. Field Robot. **25**(9), 674–724 (2008). Wiley InterScience

Reichart, G., Bielefeld, J.: Einflüsse von Fahrerassistenzsystemen auf die Systemarchitektur im Kraftfahrzeug. In: Winner, H., Hakuli, S., Wolf, G. (eds.) Handbuch Fahrerassistenzsysteme. 2nd edn. Vieweg/Teubner, Wiesbaden (2012)

Rosenblatt, J.: DAMN: a distributed architecture for mobile navigation. J. Exp. Theor. Artif. Intell. **9**(2–3), 339–360 (1997)

Kortenkamp, D., Simmons, R.: Robotic system architectures and programming. In: Siciliano, B., Khatib, O. (eds.) Springer Handbook of Robotics, Springer, Berlin (2008)

Schreier, M., Willert, V.: Robust free space detection in occupancy grid maps by methods of image analysis and dynamic B-Spline contour tracking. In: IEEE Conference on Intelligent Transportation Systems, Anchorage (2012)

Simmons, R.: Structured control for autonomous robots. In: IEEE Transactions on Robotics and Automation, pp. 10–1 (1994)

Starke, G.: Effektive Software-Architekturen, 4th edn. Hanser, München (2009)

Tölle, W.: Ein Fahrmanöverkonzept für einen maschinellen Kopiloten. VDI, Düsseldorf (1996)

Vogel, O., Mehling, U., Neumann, T., Thomas, A., Chughtai, A., Völter, M., Zdun, U.: Software-Architektur. Spektrum Akademischer, Heidelberg (2005)

Winner, H., Weitzel, A.: Quo Vadis, FAS?. In: Winner, H., Hakuli, S., Wolf, G. (eds.) Handbuch Fahrerassistenzsysteme. 2nd edn. Vieweg/Teubner, Wiesbaden (2012)

Zeng, H.: HAVEit—A Driver Centric Vehicle Automation System with a Scalable and Flexible Architecture, 19. Aachener Kolloquium Fahrzeug- und Motorentechnik, Aachen (2010)

Chapter 4
Requirements Analysis for a Universal System Architecture for Ecological and Economical Driver Assistance Systems

Peter Korzenietz

4.1 Introduction

4.1.1 Motivation

Advanced Driver Assistance Systems (ADAS) are becoming more and more relevant both for drivers and for the automotive industry. The availability and acceptance of ADAS are increasing, as can be seen in the growing diffusion of systems available in all different vehicle segments. Comfort systems in the past only available in the luxury segment, for example Adaptive Cruise Control (ACC), can now be ordered in vehicles in the compact segment. Even in the subcompact segment first advanced safety features, like an emergency brake for urban environment, can be found (Ruholl 2011, p. 10). Besides the growing spread of ADAS over all types of vehicles, the density of systems within single vehicles is increasing as well. Comfort and safety systems can be ordered and combined in a wide range.

This leads to greater complexity in multiple aspects. The more capable the systems get, the more main and sub functions are included which have to work together in a coordinated and safe way. This complexity inside the systems leads automatically to a higher effort in the development process. When a specific function of a complex system is being changed, this may affect may other functions of the system and thus the system as a whole.

Along with the interaction of functions inside a system, the systems themselves start to interact with each other. A complex network of functions across the whole vehicle is the consequence.

P. Korzenietz (✉)
Fachgebiet Fahrzeugtechnik, Technische Universität Darmstadt,
Darmstadt, Germany
e-mail: korzenietz@fzd.tu-darmstadt.de

M. Maurer and H. Winner (eds.), *Automotive Systems Engineering*,
DOI: 10.1007/978-3-642-36455-6_4,

The next big step after networking within the vehicle itself is networking with the environment. The vehicle can be connected to the internet, to the infrastructure and to other vehicles, e.g. via Vehicle2X (V2X) communication.

Together with these trends, eco friendliness is the other big trend in automotive engineering. In this context especially, higher energy efficiency is a relevant dimension for the customer on one hand and the automotive industry on the other. An energy-efficient vehicle has two main advantages for the customer. With fuel becoming more and more expensive, an energy-efficient car can help to reduce fuel costs. This is relevant for vehicles with an internal combustion engine. The second advantage of an energy-efficient vehicle is the increased range. This point is especially relevant to Electric Vehicles (EV) in the light of the current limited capacity of used batteries. Overall, the customer can benefit in terms of economy and comfort. The car manufacturers benefit from efficient vehicles because they adhere to statutory requirements. To match the current and future emission regulations, greater energy efficiency is part of the right path.

A major trend in research and development is to bring together these two trends (increasing networking in- and outside the vehicle and the need to save energy) and to develop vehicular systems that can raise energy efficiency by linking different domains and systems of a vehicle. Based on these linked domains, new networked ecological functionalities can be developed. In this chapter, the term *eco* (as in *eco functionalities*) denotes ecological.

4.1.2 State of the Art

The concept of using networking to enable new eco functionalities is a subject investigated in research. In literature, different approaches can be found, for example (Neunzig 2003; Dorrer 2004; Wilde 2009). Some of the approaches follow the target of raising efficiency by optimizing the longitudinal movement of the vehicle. An example is an extended ACC that is able to adjust the velocity not only in line with the vehicle in front but also according to road geometry and speed regulations. With an extended preview horizon based on a precise digital map, the driving strategy can be optimized according to efficiency criteria (Roth et al. 2011, p. 1453–1467). While a typical ACC is already a networked system, the integration and processing of additional information (like preview data from digital maps or V2X communication) extends the degree of networking even more.

In different research projects, system architectures are being derived with the target of structuring the systems in different layers or modules. In the project eFuture, the system architecture is divided into four layers: perception, command, execution and energy (Scheuch 2011, p. 29). While the proposed layer classification may be universally applicable, the functional focus is set on optimizing the range and safety of electric vehicles. The Active Green Driving system in the HAVEit project, however, intends to increase the operating mode efficiency of hybrid power trains and to generate driving recommendations for the driver (Sanfridson et al. 2011, p. 17).

The automated execution of the optimized speed profiles is not considered. The functional modules, however, are again assigned to perception, command and execution layers and driver interface components. A more detailed approach to a modular arrangement of components for an energy-efficient driver assistance system is proposed by Themann and Eckstein (2012, p. 1024). Together with a module for describing the driving environment based on different sensors and cooperative technologies, explicit driver and vehicle models are part of the system architecture. This allows the direct adjustment of the optimization strategy to a given driver and vehicle type. But the assistance strategy only considers the optimization of the driving mode and not of the operating mode as in hybrid vehicles. A similar modular approach is described in the extended ACC by Porsche (Porsche InnoDrive) (Roth et al. 2011, p. 1453–1467).

The analysis of the system architectures mentioned in research projects above shows that both layered and modular system architectures are being used for energy-efficient driver assistance systems. Nevertheless, both types of architectural approaches have in common that the solutions obtained are characterized by a fixed and predefined set of eco functionalities. Furthermore, some of the introduced systems only focus on selected vehicle or propulsion types (conventional, hybrid or electric vehicles) with type-given constraints. A predefined set of eco functionalities and focusing on selected vehicle or propulsion types lead to system architectures that are not robust towards changes. An easy integration of new functionalities or the adaption of the system to other vehicle concepts cannot be ensured.

4.1.3 Targets

The eco$_2$DAS research project addresses these deficits. The target is to develop a system architecture that is universally applicable to all current and foreseeable vehicle/propulsion types in present-day and future passenger cars and supports each possible set of eco functionalities.

For developers, the eco$_2$DAS system architecture is meant to act as a pre-defined framework. A predefined framework offers the advantage that the development process can be shortened and made more efficient, because the development focus can be set directly on function level instead of first defining the system level. Since the system architecture is the same for each eco function, the complexity is reduced, because additional functions, changes in functions and the transfer to other vehicle types does not require a change in the system level.

The basic approach for the design of the eco$_2$DAS system architecture is to enable an interface structure between all function units in the vehicle, such as ADAS, drive train, engine, infotainment systems and thermal managementation (Fig. 4.1). Based on this network, eco functions have access to a variety of information, allowing a holistic energetic optimization of the entire vehicle.

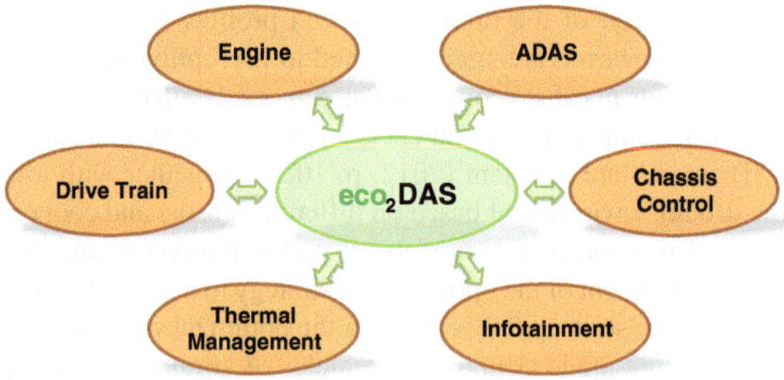

Fig. 4.1 Networked system of function units

The basic aspects in developing the system architecture are the definiton of communication signals, interfaces for information exchange and core modules on the system level.

4.1.4 eco$_2$DAS in the Context of Automotive Systems Engineering

Systems engineering means the successful development of systems (Weilkiens 2008, p. 11). The description of the structure of a system, its components and the visible properties and relations of the components to each other is fulfilled by a system architecture (Vogel 2005, p. 48). Therefore, the development of a system architecture can be seen as a part of systems engineering.

The challenge in designing the eco$_2$DAS system architecture is the high degree of aspired connectivity. The large quantity of connections and shared information between the different vehicle function units increases the complexity of whole the network significantly. This is already the case when assuming just a defined set of functions and given vehicle constraints (esp. drive train and ADAS). If an even wider range of possible functions and different vehicle constraints has to be considered, as in the eco$_2$DAS system architecture, then the potential complexity grows even more. For example, for an electric vehicle with limited energy storage, an efficient driving mode with a low acceleration profile and moderate use of other electric consumers like air conditioning has a far greater influence on the available cruising range than it has in conventional or hybrid vehicles. If battery loading time (and the resulting discomfort for the driver) is covered as well, the optimization problem gets even more complex and interconnected. The eco$_2$DAS system architecture intends to be able to carry out such complex optimization functions with its high amount of links.

To manage this challenge, a structured, top-down approach is proposed. Systems engineering can deliver the necessary methods to reach this target.

4.1.5 Examined System Architecture Level and View

Architectures can be described in different levels, which represent different degrees of abstraction (Vogel 2005, p. 66) and address level-specific problems (Vogel 2005, p. 68). Vogel (2005, p. 67) mentions the organization, the system and the component levels. The organization level deals with the complete organization and the resulting specifications for the systems within the organization. On the system level, the specific systems are the topic of interest, including their functionalities and interfaces to other systems. Finally, on the component level, the particular components of the systems and their structure are defined. According to this description, the eco$_2$DAS system architecture is settled mainly on the component level, because the architecture aims to declare the functional components, their structure and the interfaces between the components. Nevertheless, with the aspired high degree of connectivity to other systems, like other vehicles or traffic management service providers, the system level has to be taken into account. Special consideration has to be given to the interfaces with their possible information exchange.

Different architecture views on a system can be provided for each level, with each architecture view representing a different stakeholder's perspective on the same system. The aim is to clearly describe a complex system according to the needed purpose (Vogel 2005, p. 76). In this chapter the focus is set on the view that describes the components and their relation to each other. Example views that deal with these aspects are the component view (Starke 2005, p. 85) or the logic view (Vogel 2005, p. 80). In the automotive architecture model "AutoMoDe" this view (or abstraction level) is called Functional Analysis Architecture (FAA). The FAA aims to describe the vehicular functionalities, their structure and dependencies. The functionalities are described independently in terms of implementation in software and hardware (Bauer et al. 2007, p. 48).

4.2 Derivation of Requirements

In a typical development process for electronic systems, user requirements are the basis for defining the logical system requirements (Schäuffele and Zurawka 2006, p. 152). While user requirements represent the users' view on a system/product, the logical system requirements represent the view of the technical disciplines responsible for system development. According to Schäuffele and Zurawka (2006, p. 152), the logical system requirements describe those properties a system has to have and those it must not have. Another view is the differentiation between functional and nonfunctional requirements. While functional requirements describe the normal operation and malfunctions, nonfunctional requirements describe the remaining requirements, such as given constraints. The outcome of this process is a logical system architecture that describes the functional relations.

The major difference in defining the requirements for the eco$_2$DAS system architecture is the absence of user requirements and logical system requirements that address defined functions of a system. The approach is not to define what the eco$_2$DAS system should do and what functions follow from this. Rather, instead of providing a predefined set of functions, the eco$_2$DAS system architecture aims to provide a framework for many possible eco functions. Hence, in this context, requirements are needed that do not define what a system should do but that describe what the system architecture has to perform.

4.2.1 Approach for Requirements Analysis

In the classic development process, user requirements based on use cases are the basis for the system requirements. A use case can be described as a list of actions between a system and a role, which lead to observable results that have some useful outcome (Weilkiens 2008, p. 231). For the development of the eco$_2$DAS system architecture, such use cases that describe in a precise way the interaction between the driver and the system are not available. But nevertheless, an information basis as a starting point for the further development is necessary. To provide such an information basis, basic use scenarios have to be derived that describe possible interactions between eco assisting ADAS, the driver and the vehicle. In that sense, the basic use scenarios are comparable to use cases, but they describe the interaction in a broader but less precise manner so as to obtain an overview of possible system functionalities.

To obtain the basic use scenarios, the approach shown in Fig. 4.2 is proposed. First, a driver-vehicle-environment model is given as a classification approach to describe possible interactions. Based on this model, an influence analysis is conducted with the target to identify factors that influence energy consumption. After that, the basic use scenarios are defined. Finally, based on the scenarios, the functional requirements are derived.

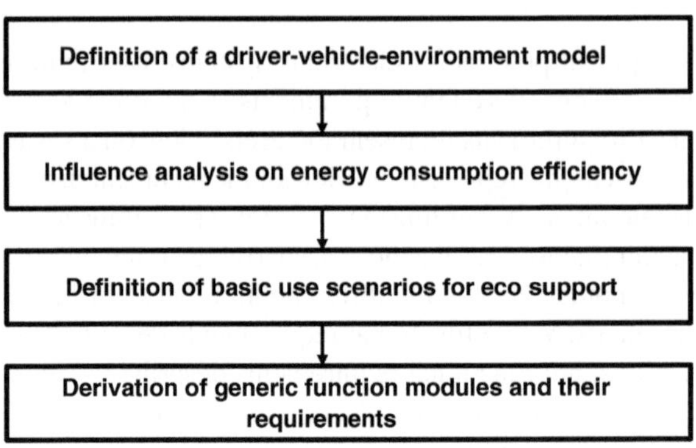

Fig. 4.2 Approach for requirements analysis

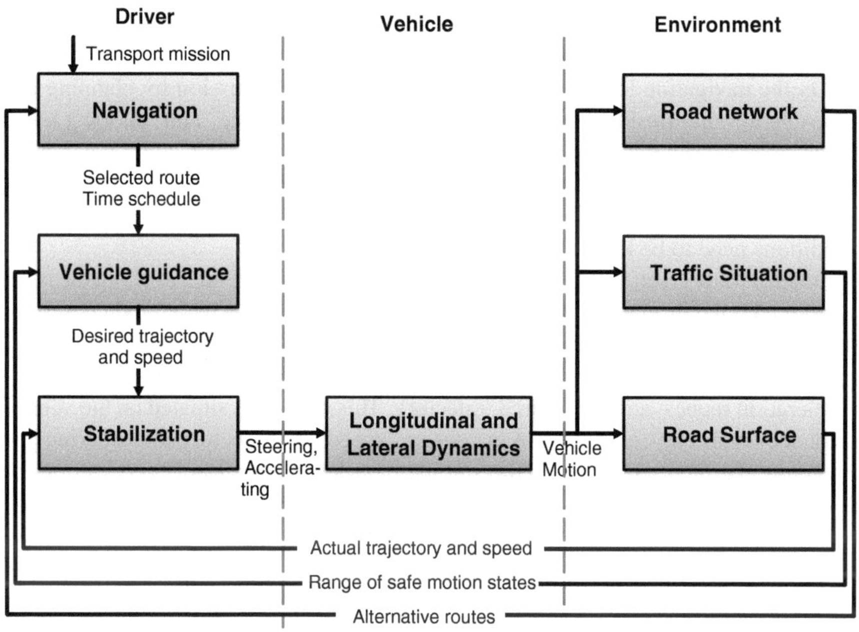

Fig. 4.3 Three-level hierarchy of driving task (Donges 1982, p.183–190)

4.2.2 Definition of a Driver-Vehicle-Environment Model

An established model to describe the interaction between driver, vehicle and environment is the three-level hierarchy of driving task according to Donges (1982, p. 183–190) (Fig. 4.3). The highest level is the navigation layer, on with the route is planned. On the second layer (vehicle guidance layer) the chosen route is followed and on the third layer (stabilization layer) the required inputs for the vehicle are generated (acceleration and steering). In traditional vehicle guidance the driver acts as a controller.

4.2.3 Influence Analysis on Energy Consumption Efficiency

Based on the three-level hierarchy, an analysis is made of how the driver, the vehicle and the environment can affect energy consumption.

4.2.3.1 Driver

The energy consumption efficiency of a transport mission is mainly influenced by the behavior of the driver (Dorrer 2004, p. 35) on the navigation and vehicle guidance levels according to Neunzig (2003, p. 20). For a more precise description, the stabilization level has to be considered as well.

On the navigation level the driver affects the transport mission by planning and choosing a route. Besides the criteria "shortest travel distance" and "shortest travel time" s/he also can take "lowest fuel/energy consumption" into account when planning and choosing a route. To calculate the energy demand of a route, several aspects need to be considered, especially the topography, the signage, the weather and traffic conditions need to be known and predicted.

When a route is set, the driver can affect the energy demand for travelling the route by adjusting the driving mode and the operating mode. The driving mode describes the vehicle guidance dependent on the curves on the road and road conditions, other traffic participants and the traffic regulation e.g. by signs. The output is a velocity/acceleration request and the desired trajectory. This process is situated on the vehicle guidance level. The conversion of the desired speed and trajectory into steering and throttle/brake inputs is done on the stabilization level. This whole process can be described as the primary driving task according to Bubb (2002, p. 22–23). This use of the steering and throttle/brake on basis of the desired trajectory and speed is part of the operating mode (Neunzig 2003, p. 21). But the operating mode consists of more components. The operation of the gearbox, the turn signals and other functions like the wipers or the light switch are also part of the operating mode according to Neunzig. Bubb (2002, p. 22–23) summarizes these additional tasks as the secondary driving tasks. They need to be done when driving a vehicle, but they are not essential for keeping the vehicle on the road. Tasks that are not relevant for the actual driving task but for comfort and entertainment are called tertiary tasks. Such tasks are the operation of the air conditioning and radio. The secondary and tertiary tasks are not part of the conventional three-level hierarchy. For better comprehension of the driving and operating mode, the three-level hierarchy is completed by the secondary and tertiary driving tasks and looks as shown in Fig. 4.4.

The driving mode represents the desired speed/acceleration profile of the driver and thus the energy demand of a vehicle. If a driver wants to accelerate quickly and travel at a high speed, the result is a higher energy demand compared to a moderate driving mode. The satisfaction of this demand is realized in the operating mode, because by operating the vehicle the desired speed/acceleration profile can be transferred into vehicle motion. The operating mode is not necessarily directly linked to a given driving mode. With a gearbox, for example, several gears can be chosen for a given speed. With a continuously variable transmission (CVT), the amount of combinations is even higher and in a hybrid vehicle with a second propulsion system the operating mode is completely independent of the driving mode. Both aspects can be optimized separately in terms of energy consumption. A system can support the optimization, for example by recommending an efficient travel velocity (driving mode) or by recommending an efficient gear or moderate passenger compartment temperature (operating mode). In order to maximize energy efficiency gain, both modes need to be treated together.

With his/her behavior in form of planning the route, and driving and operating the vehicle, the driver can directly affect the energy consumption for a given transport

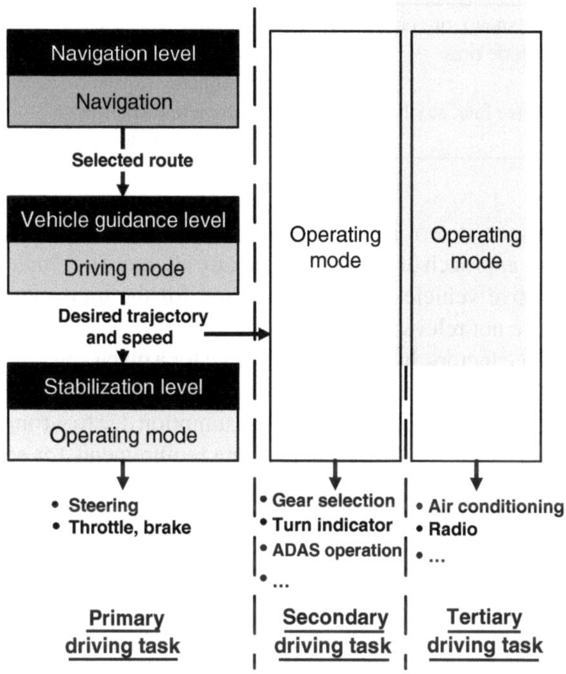

Fig. 4.4 Extended three-level hierarchy driver model

mission. If these driving tasks can be optimized in terms of energy efficiency, the energy demand for a transport mission can be reduced.

4.2.3.2 Vehicle

Influencing factors on energy consumption from the vehicle side can be divided into factors from the manufacturer and factors from the vehicle user. According to Dorrer (2004, p. 34), Table 4.1 lists the influencing factors from the manufacturers and Table 4.2 lists factors from vehicle users:

Table 4.1 Manufacturer-affected influence factors on energy consumption (Dorrer 2004, p. 34)

Influence factor	Influenced variable
Vehicle mass (empty), wheel/tire type	Rolling resistance
Body design and size, wheel/tire form	Air drag
Vehicle mass	Climbing resistance, acceleration resistance
Drive line, auxiliary units	Drive line efficiency
Engine (type, parameters, combustion process)	Engine efficiency

Table 4.2 Vehicle-user-affected influence factors on energy consumption (Dorrer 2004, p. 34)

Influence factor	Influenced variable
Payload, tire type (e.g. snow), tire pressure	Rolling resistance
Roof extension, wide-base tires	Air drag
Payload	Climbing resistance, acceleration resistance
Technical condition drive line, auxiliary units	Drive line efficiency
Engine maintenance	Engine efficiency

The influence factors determined by the manufactures are fixed for a given vehicle. Since the eco$_2$DAS approach does not consider any alternative vehicle than the given car (i.e. no alternative vehicle will be considered for the transport mission) these influence factors are not relevant.

The user influence factors, however, can change for a given vehicle. Since possible efficiency optimization functions may work with those factors (e.g. information to the driver about the expected additional fuel consumption due to wrong tire pressure), they have to be taken into account when deriving requirements for eco systems.

4.2.3.3 Environment

Influence factors given by the environment are shown in Table 4.3.

These influence factors cannot be changed with assistance systems in a vehicle, but route planning and choosing functions/algorithms can use the factors for calculating an energy efficient route.

4.2.3.4 Breakdown of a Transport Mission

To be able to energetically optimize a complete transport mission it can be divided into a *trip view* and a *maneuver view*. Both views differ in their temporal dimension and prediction horizon and, therefore, in their possible approach for optimization.

The *trip view* spans the entire transport mission from the starting point to the destination. Based on these two points, the energetically optimized routing and route

Table 4.3 Environment-affected influence factors on energy consumption (Dorrer 2004, p. 34)

Influence factor	Influenced variable
Road surfacing, routing/curves, weather conditions	Rolling resistance
Track topography	Climbing resistance
Speed regulations, traffic and weather conditions, curves	Travel speed
Traffic dynamics	Acceleration
Routing and route length	Travel distance
Traffic situation (congestions), traffic lights	Travel time
Motor operation points (dependent on speed, traffic), temperature, elevation	Engine efficiency

guiding can be generated. This can be done by predicting ecological and economic costs on base of ego vehicle properties, road network properties and traffic situation prediction. The view on the road and traffic state of a route segment is thereby a macroscopic view, based on the stochastic description of the traffic constraints, because the long range prediction e.g. of traffic-light states or the speed profiles of individual vehicles seems neither beneficial nor feasible. Usually, the route is planned before the trip and can be adapted while already travelling. Nevertheless, in addition to a route recommendation, the trip view can contain a speed profile as an outcome that can be adapted to local circumstances as described in the maneuver view.

The *maneuver view*, however, does not span the entire transport mission but rather the current and upcoming maneuvers. The length of a maneuver is influenced by traffic and road constraints. A typical maneuver can be an approach to a red traffic light or a follow-up to a vehicle in front. On this level, the exact current and predicted states of the road network (e.g. traffic light) and the surrounding vehicles (positions and speeds) need to be known because this information is the basis for generating driving and operating strategies like speed profiles or hybrid operating modes. In comparison to the trip view, the maneuver view needs a shorter preview horizon but a more detailed local traffic environment description.

4.2.3.5 Review of Influencing Factors

The main influence variables on energy consumption are the driving resistances which, for their part depend on influence factors (for example mass/payload) and other variables (for example speed and acceleration). For a given vehicle, the influence factors resulting from the vehicle are either fixed, like body design or empty mass, or can be optimized by the vehicle user, like adjusting the right tire pressure or dismounting roof extensions if not needed. For these measures, simple advice to the driver seems sufficient.

The influence factors resulting from the environment (e.g. topography or traffic situation), however, are variable and determine the space in which the transport mission can be executed (e.g. highway vs. expressway) and the ego driving behavior (e.g. free driving vs. stop-and-go traffic). In this surrounding, the task for the driver is to consider all the relevant influencing factors and create an efficient route on the trip level and efficient trajectories (speed/acceleration profiles) on the maneuver level. With route and trajectory choices, the driver determines the driving resistances and therefore the energy demand for a transport mission. The more information about the environment is available, the more the transport mission can be planned and executed in line with efficiency aspects. Against this background, it seems meaningful to assist the driver in his/her perception of the environment factors and in generating driving strategies.

Other influence variables on energy consumption are the efficiencies of the drive line, especially when it contains an internal combustion engine. The fuel efficiency is dependent on the load and the engine speed. For this reason, assistance functions are promising that support the driver in generating and executing operating modes that result in efficient drive train/engine operating points. These functions can handle simple gear recommendation or complex hybrid operating strategies.

4.2.4 *Definition of Basic Use Scenarios for Eco Support*

Based on the influence analysis conducted, basic use scenarios can be formulated. A basic use scenario describes in which manner an ecological driver assistance system may interact with the driver and/or the vehicle. In line with the classification introduced in Sect. 4.2.3.1 the use scenarios are classified as navigation, driving mode and operating mode (Table 4.4).

The basic use scenarios give an overview of the interaction between the system and the driver. In that context they represent the user's view on the system. A recommendation on speed, for example, is settled in the driving mode for the driver, but the function for generating the recommended velocity profile can be settled in the navigation and route planning algorithms. In this case, the function delivers a route recommendation and a speed profile for the entire route.

4.3 Functional Modules and Their Requirements

In this chapter, focus is set on the Functional Analysis Architecture (FAA) which describes the system functionalities, their structure and the information flows/interfaces (see Sect. 4.1.5). This view is the basis for the further concretization of the system regarding its implementation as soft-and hardware, which results in additional architecture that will not be considered here.

The following requirements describe the functionalities on the FAA level. Functionalities mean elementary and user-visible functions that are needed to provide the intended eco-assistance for the driver. This means that the description of the FAA level concentrates on core components of the system application but not on the subordinated functions such as diagnosis or system monitoring (Bauer et al. 2007, p. 48).

Figure 4.5 gives an overview of the function modules and their structure. In the following a description of the modules and their requirements is given.

Energetic Optimization

This functionality is the core module for achieving the aspired higher efficiency because this is the component where the energetic optimization strategy is developed. The optimization strategy can relate to different transport mission horizons, which means the optimization can involve the entire trip, the upcoming maneuvers or a combination of both. The outcome can contain energetically optimized route recommendations and guidance, driving profiles and operating strategies for the vehicular function units. The optimization module of the system architecture is not necessarily

Table 4.4 Basic use scenarios for ecological support functionalities

Classification	Basic use scenario	Description
Navigation	**Route recommendation** (pre-trip)	System: Provides eco relevant information (e.g. traffic situation, topography, weather) for the current time, adds forecast information (Preview horizon: Complete trip/several hours), plans eco route with alternatives and recommends a route Driver: Chooses one route *Example: The system recommends two energy-efficient routes that differ in energy and time demand*
	Route guidance + **Route adaption** (on-trip)	System: Guides the driver along the chosen eco-route and adapts the route if necessary due to irregularities Driver: Drives the vehicle (incl. planning of driving and operating mode) *Example: The system adapts the route due to expected stop&go traffic and recommends an alternative route*
Driving mode	**Information supply**	System: Provides eco-relevant information (e.g. traffic, topography, traffic-light status) for the current and next relevant driving situations (Preview horizon: next maneuvers/several minutes) Driver: Plans driving mode and drives vehicle *Example: The system displays that the approaching traffic light will turn red in a few seconds*
	Driving mode recommendation	System: Plans an efficient driving mode for the next maneuver and recommends it to the driver. Driver: Operates the vehicle according to the driving mode recommendation *Example: The system recommends lifting the foot from the accelerator pedal when approaching a curve without braking would be possible*
	Automated maneuver control	System: Plans an efficient driving mode for the next (longitudinal) maneuver for the complete vehicle and conducts the maneuver (system is responsible for primary driving task (throttle, brake)) Driver: Monitors the execution of the maneuver and operates functions from the secondary driving task (especially operation of the gearbox) *Example: The system is responsible for the longitudinal control (e.g. ACC) and the driver operates a manual gearbox* Automated maneuver control implies that the system is responsible for both driving and operating mode. If only the driving mode is generated by the system, it is the use case "Driving mode recommendation".
Operating mode	**Information supply**	System: Provides eco-relevant vehicle-status information (e.g. efficiency of current motor operating point) Driver: Plans operating mode and controls driving variables *Example: The system indicates the load of the air conditioning*
	Operating mode recommendation	System: Generates an efficient operating mode for observed function units (e.g. engine, gearbox, air conditioning) Driver: Operates actuating variables according to the operating mode recommendation. *Example: The system recommends turning off the air conditioning because the trip will soon end*
	Assisted Driving	System: Generates an efficient operating mode for observed function units and operates the units Driver: Monitors the execution of the operating mode *Example: The system uses the electric motor of a plug-in hybrid vehicle for the last part of a trip, so that the batteries will be empty when arriving home.*

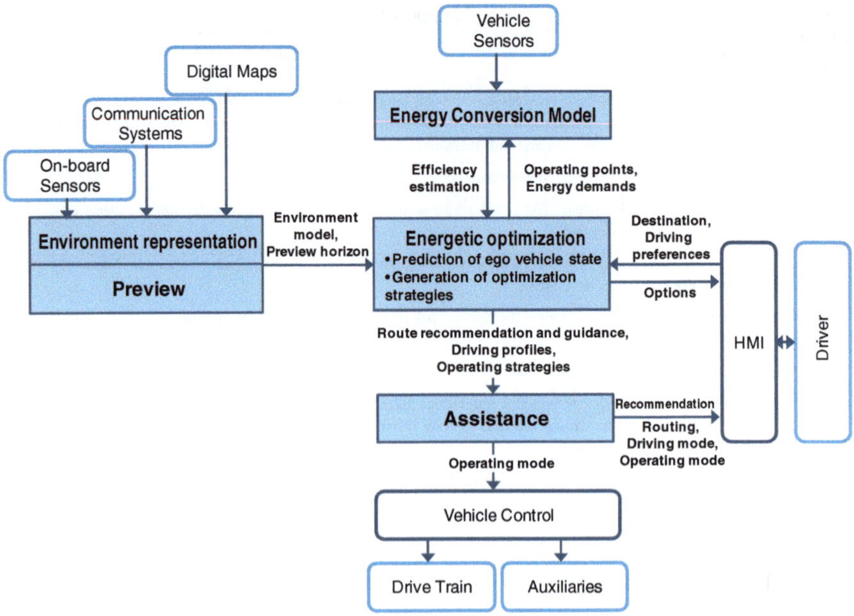

Fig. 4.5 Function overview based on the Functional Analysis Architecture; HMI: Human Machine Interface

located on a control unit in the vehicle. It could be envisioned to execute the optimization especially for the entire trip on an external computer system (cloud-based service) so that the optimization task would not be limited by the comparatively low computing power of vehicular control units.

Based on preview functions, energy consumption can be reduced further (Zlocki 2010, p. 65). With preview information, for example, it is possible to optimize the energy management in hybrid drive trains or the longitudinal vehicle control according to the previewed route properties and the traffic situation. In other words, the preview contains the future constraints which need to be considered for optimization. For this reason, energetic optimization needs as input an environment representation with preview.

Part of the energetic optimization is the prediction of the ego vehicle state. Based on the driving-environment representation and preview, the upcoming driving resistances and energy demands can be estimated. In combination with an energy conversion model of the vehicle that is provided as an additional input, the prediction unit can determine the efficiency of routing, driving and operating strategy. The prediction horizon can cover the next maneuvers (e.g. approaching a traffic light) or can extend up to the entire trip. Necessary inputs, therefore, are the destination of the trip or at least the most probable path, a global digital map with attributes and the traffic environment forecast.

Environment Representation and Preview

To enhance the energetic optimization, the traffic environment state and the preview have to be considered. These contain static information like topography, road layout and signage and dynamic information like speeds/accelerations of the surrounding vehicles, traffic density or traffic-light states. While the static information just requires a spatial preview (e.g. with a digital map and the planned route) because topography and road layout do not change in terms of time, dynamic information, however, needs a temporal preview. The traffic density or the weather, for example, are highly dependent on the time of day. In line with the breakdown of a transport mission, this module contains the representation and preview for the maneuver and trip views.

For the perception of the environment, different sources are available. They vary in technological approach, range or information provided. While on-board sensors such as radar, lidar or cameras deliver a picture of the surrounding traffic situation with a high actuality but limited range, communication-based sensor systems like Vehicle2Vehicle/Vehicle2Infrastrucutre (V2X) or extended floating car data (XFCD) offer a longer range, but with higher uncertainty due to the longer prediction horizon. Since the target of energetic optimization is to optimize the complete transport mission on each level (from each maneuver to the entire trip) the information demand on the environment reaches from the short-range surrounding traffic situation through to the global road network state with its attributes. To obtain a homogeneous environment model an information fusion is necessary. The environment representation and preview component is responsible for combining the different information and then providing it to the energetic optimization module.

Energy Conversion Model

The energy conversion model is the basis for predicting the future energy flow of the vehicle. With operating points and energetic demand as inputs, it is possible to determine the expected energy flow and, therefore, the efficiency for a maneuver or a trip.

Since the energy conversion model is a separate component in the system architecture, it should be modeled for different vehicle types without the need to change the energetic optimization module itself.

Assistance

The output of the energetic optimization consists of route recommendations and guidance, driving profiles and operating strategies. The underlying assistance concept allows these outcomes to either be communicated to driver as recommendations or executed by the vehicle itself. In the first case, a recommendation interface has to be able to communicate with different HMI concepts like displays or active gas pedals. In the second case, an execution interface is necessary that connects the drive train or hybrid controls and the energy and battery management controls to the system. The execution interface has to be able to support all different degrees of automation, beginning, for example, from air conditioning or gearbox control up to a fully autonomous vehicle.

In this context, the concept of an arbitration unit is useful. The arbitration unit decides which system has the highest priority to affect the related controls. In a critical driving situation, for example, emergency brake commands have a higher priority than eco systems. In normal driving situations, however, the eco system might be the leading system.

It is important to mention that the system architecture and the module requirements describe a generic system design that does not match a pre-defined set of functions. The purpose of the eco$_2$DAS architecture is to act as framework, which can be instantiated for defined eco functions. For example, one possible eco function could aim to optimize the operating mode of a hybrid vehicle dependent on the upcoming topography, while another function recommends driving strategies for stop-and-go traffic situations. Both functions may differ in their optimization strategy and their demand on environment representation but the requirement is that both function instances have to work on the basis of the eco$_2$DAS system architecture.

4.4 Conclusion and Outlook

This chapter proposes an approach for defining requirements for a universal system architecture for ecological and economical driver assistance systems. Based on the three-level hierarchy model and an influence analysis, basic use scenarios are pictured that give an overview of possible assistance functionalities for the driver. The proposed functionalities provide assistance in generating energy-efficient routes, driving profiles and operating strategies. These functionalities are the basis on which to derive requirements for the desired Functional Analysis Architecture. The main requirement is the availability of an energetic optimization module that contains the assistance functionalities for energy efficiency gain and represents the core module in the system architecture. The second main requirement is the availability of an environment representation and preview component that provides spatial and temporal information about the environment in terms of road and traffic states for the energetic optimization. In combination with an energy conversion model of the vehicle and a recommendation and execution component, the requirements deliver a first overview of the functional analysis architecture.

The upcoming development steps will be to further concretize the mentioned requirements. The sub functions and interfaces of the introduced functional units will be specified. Especially the energetic optimization sub functions such as the prediction of ego vehicle energy consumption and the optimization function itself need to be specified. The big challenge is to ensure that the resulting system architecture stays universally applicable. If just one of the relevant assistance functions should not be implementable within the future eco$_2$DAS system architecture, the main requirement that the system architecture be universally applicable is not met.

After the Functional Analysis Architecture is specified according to the AutoMoDe level model, the Functional Design Architecture (FDA) will be developed.

The FDA will contain the description of the software view of the system and be implementation oriented.

References

Bauer, A., Broy, M., Romberg, J., Schätz, B., Braun, P., Freund, U., Mata, N., Sandner, R., Mai, P., Ziegenbein, D.: Das AutoMoDe-Projekt—Modellbasierte Entwicklung softwareintensiver Systeme im Automobil, Informatik—Forschung und Entwicklung Band 22, pp. 45–57, December 2007, Springer Verlag Heidelberg (2007)

Bubb, H.: Der Fahrprozess—Informationsverarbeitung durch den Fahrer, VDA 4. Technischer Kongress "Sicherheit durch Elektronik", March 20–21. Germany, Stuttgart (2002)

Donges, E.: Aspekte der Aktiven Sicherheit bei der Führung von Personenkraftwagen, Automobil-Industrie 2/82, pp. 183–190, Vogel Business Media GmbH & Co. KG Würzburg (1982)

Dorrer, C.: Effizienzbestimmung von Fahrweisen und Fahrerassistenz zur Reduzierung des Kraftstoffverbrauchs unter Nutzung telematischer Informationen. Schriftenreihe des Instituts für Verbrennungsmotoren und Kraftfahrwesen der Universität Stuttgart, Expert Verlag Renningen (2004)

Neunzig, D.: Fahrerassistenzsysteme zur Verbrauchsminderung von Kraftfahrzeugen—Anforderungen, Realisierung und Bewertung, D82 (Dissertation RWTH Aachen), Forschungsgesellschaft Kraftfahrwesen Aachen mbh (fka) Aachen (2003)

Roth, M., Radke, T., Lederer, M., Gauterin, F., Frey, M., Steinbrecher, C., Schröter, J., Goslar, M.: Porsche InnoDrive - An Innovative Approach for the Future of Driving, 20th Aachen Colloquium Automobile and Engine Technology, October 8–10. Aachen, Germany (2011)

Ruholl, H.: Vom Showcar zur Serie, ATZextra - Der neue VW up!, pp. 8–10, Springer Vieweg Verlag Wiesbaden (2011)

Sanfridson, M., Lundgren A., Jarlengrip, J., Grubb G., Feng, L., Bruce, M.: Deliverable D54.3—Active Green Driving: 1st System Functionality, HAVEit project, (2011) http://www.haveit-eu.org/LH2Uploads/ItemsContent/24/HAVEit_212154_D54.3_Public.pdf

Schäuffele, J., Zurawka, T.: Automotive Software Engineering. Grundlagen, Prozesse, Methoden und Werkzeuge effizient einsetzen, Vieweg Verlag Wiesbaden (2006)

Scheuch, V.: E/E-Architekturen für batterie-elektrische Fahrzeuge, ATZ elektronik 06/2011, pp. 28–33. Springer Vieweg Verlag Wiesbaden (2011)

Starke, G.: Effektive Software-Architekturen. Ein praktischer Leitfaden, Hanser Verlag München (2005)

Themann, P., Eckstein, L.: Modular Approach To Energy Efficient Driver Assistance Incorporating Driver Acceptance, IEEE Intelligent Vehicles Symposium 2012, June 3–7. Alcalá de Henares, Spain (2012)

Vogel, O.: Software-Architektur. Grundlagen-Konzepte-Praxis, Spektrum Akademischer Verlag Heidelberg (2005)

Weilkiens, T.: Systems Engineering mit SysML/UML, dpunkt Verlag Heidelberg (2008)

Wilde, A.: Eine modulare Funktionsarchitektur für adaptives und vorausschauendes Energiemanagement in Hybridfahrzeugen. Dissertation Technische Universität München (2009)

Zlocki, A.: Fahrzeuglängsregelung mit kartenbasierter Vorausschau, D82 (Dissertation RWTH Aachen), Forschungsgesellschaft Kraftfahrwesen Aachen mbh (fka) Aachen (2010)

Chapter 5
Static Software Architecture of the Sensor Data Fusion Module of the Stadtpilot Project

Sebastian Ohl

5.1 Introduction

Driving automatically on inner-city roads has become more and more popular in recent years. After simulating inner-city traffic at the DARPA Urban Challenge, some groups have moved on and successfully mastered the challenges of real-world inner-city traffic. At TU Braunschweig, the Stadtpilot project started in summer 2008 to master automatic driving on the inner-city ring road of Braunschweig. As depicted in (Saust et al. 2011), the experimental vehicle successfully passed many test runs with autonomous longitudinal and lateral guidance on Braunschweig's city ring road in autumn 2010. In this set-up, the velocity of the experimental vehicle reaches up to 50 km/h and the number of other road users has increased significantly compared to prior projects. To sense its environment, the experimental vehicle is equipped with LIDAR and RADAR sensors for environmental perception. All sensors are fused by sensor data fusion algorithms to form a consistent view of the vehicle's surroundings. This data is passed on to the application by an uniformed interface.

Due to the fact that the Stadtpilot project has an iterative development process, the project's goals are adjusted in each iteration to realize new functionalities in the area of operation. Therefore, the requirements to the environmental perception may also change. Accordingly, the environmental perception needs to be more flexible and more adaptable than in prior projects. Furthermore, the experimental vehicle is used in projects other than Stadtpilot, e.g. project Koline, see (Saust et al. 2010). Therefore, the requirements of these applications have to be fulfilled as well.

This paper presents the static view of the sensor data fusion software architecture of the Stadtpilot project. It offers an easy to use object oriented framework for creating different sensor data fusion applications for vehicle environmental perception.

S. Ohl (✉)
Institute of Control Engineering, Technische Universität Braunschweig,
Hans-Sommer-Str. 66, D-38106 Braunschweig, Germany
e-mail: sebastian@ohl.name

M. Maurer and H. Winner (eds.), *Automotive Systems Engineering*,
DOI: 10.1007/978-3-642-36455-6_5, © Springer-Verlag Berlin Heidelberg 2013

By defining detailed interfaces between the architecture's elements down to single classes, algorithms, and processing stages can be easily replaced.

After an overview of other architectures for environmental perception, the scope of the discussed software architecture is defined. The next part introduces the global architecture of the sensor data fusion system. Afterward, the description of the interfaces for decoding a sensor's communication protocol is presented. Section 5.6 delineates the interfaces and components within the object hypotheses based sensor data fusion. This is followed by the section on the grid based sensor data fusion. Each sensor data fusion chapter is followed by an overview of the operational application. At last, some limitations of the depicted architecture are discussed and the paper is concluded by a summary.

5.2 Architectures for Environmental Perception

Papers on architecture for environmental perception or sensor data fusion architecture can be grouped into various types of publications. The first group of papers focuses on the sensors used and how they are combined to fit a specific project's need, e.g. (Al-Dhaher and Mackesy 2004; Durrant-Whyte et al. 1990; Effertz 2008). The architectures in these publications are normally limited in their usability for other projects. Another group of publications describes the structures of sensor data fusion systems in general and not in relation to a specific project. This group can be divided into older and newer papers. As the older group provides the basis of the newer papers, they will be described first. One of the first publications on the topic of data fusion is a set of slides from the year 1976 by Boyd (1976). In these sheets, the basic cycle of gathering information, processing this data to get to a decision and carrying it out is described (Boyd-cycle). In (Dasarathy 1997) and (Markin et al. 1997) the Waterfall-model[1] was published. It describes a fusion architecture as a linear process containing very abstract processing stages. Generally, these processing stages can be found in any modern sensor data fusion system. The Omnibus-model, see (Bedworth and O'Brien 1999), enhanced the prior models. The fusion process is mainly based on the model from Boyd (1976). In this model the Boyd-cycle is reinterpreted to match the needs of a sensor data fusion and not a data fusion process in general. Because of the high degree of abstraction, they do not help the developer in structuring his sensor data fusion in detail. Due to that fact, all these approaches are limited in their use. More information on these models and similar architectures can be found in (Bedworth and O'Brien 1999) or (Gad and Farooq 2002).

In the following, some newer and more practical architectures are shown. Afterward, these architectures are set in the context of this paper.

[1] This Waterfall-model in this paper referres to the model in the data fusion domain and should not be mixed up with the Waterfall-model from development process domain as introduced by Royce (1987).

5.2.1 LAAS-Architecture

Other than the architectures discussed above, the LAAS-architecture (Laboratoire d'Analyse et d'Architecture des Systèmes) describes its components and their interaction in a more detailed way. The LAAS-architecture has been developed as a common architecture for the field of robotics. It describes not only the sensor data fusion part, but also the part of creating decisions and how to carry them out. The architecture is split into five layers: physical system, logic robotics layer, functional level, execution control level, and decision level. Within every layer, specific tasks are described as well as the modules' organization. Apparently, the LAAS-architecture did not get a wider distribution because of the already existing JDL-Architecture.

5.2.2 JDL-Architecture

The JDL-architecture has been created on behalf of the US department of defense by the joint directors of laboratories' (JDL) data fusion sub-panel (according to [White 1988) as cited in (Steinberg et al. 1999)]. Figure 5.1 shows the JDL-fusion model. It is separated into five levels:

Level 0: Sub-Object Data Assessment: Level 0 is responsible for the preprocessing of signals and measurement data.
Level 1: Object Assessment: In Level 1, the preprocessed measurement data are converted into a common data format (tracks). This is performed by associating the data with measurements from previous measurement cycles followed by a tracking process.
Level 2: Situation Assessment: These tracks need to be related to other tracks and the current situation. Therefore, a situation analysis is performed in this level.
Level 3: Impact Assessment: After the situation assessment, the results need to be related to the current mission goals.

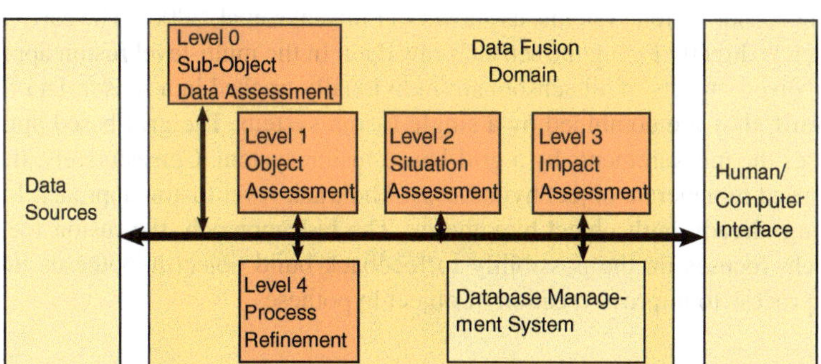

Fig. 5.1 JDL data fusion model, according to (Steinberg et al. 1999, p. 6)

Level 4: Process Refinement: In Level 4, the decisions are implemented. To improve the overall process, the algorithms in Level 4 can adapt the processes of the other levels and rearrange the system resources if needed.

As depicted in Fig. 5.1, the levels are not necessarily ordered in a hierarchical way. They are connected by shared data bus. This bus can also be connected to other infrastructure elements (e.g. databases or an HMI). By using a shared bus infrastructure, individual layers can be skipped, if they are not necessary for the specific task (Steinberg et al. 1999, p. 5).

Although developed years ago, the JDL-model forms the basis for many current sensor data fusion architectures. In the past years, it has been revised by the original authors. In (Steinberg et al. 1999), the authors rectified some misunderstandings in the usage of the original model. In (Llinas et al. 2004), some enhancements were presented. The authors proposed the removal of Level 4. Beyond that, they covered the usage of other levels in distributed systems. As a 5th optional level, an HMI-Level was introduced and added to the core fusion process. In the original model, the HMI component was drawn outside the data fusion domain.

5.2.3 ProFusion2-Architecture

An adaption of the JDL-model to the automotive sector is described in (Polychronopou los and Amditis 2006). The EU-project PReVENT developed the ProFusion2-architecture (PF2) in its sub project ProFusion. Within the PF2-architecture, three layers were defined: the Perception Layer, the Decision Application Layer, and the Action/HMI Layer. The Perception Layer combines Layer 0 and 1 of the JDL-architecture. Its Layers 2 and 3 can be found in the Decision Application Layer. The Action/HMI Layer represents Layer 5 of the JDL-architecture. Within the Perception Layer, the JDL-Layer 1 is separated into three individual components with different abstraction levels (Feature, Track, and Object). This categorization is derived from the extended 4D environment model described in (Scheunert et al. 2006). Within the PReVENT-Project, the PF2-architecture has been used for five different sensor fusion systems, as shown in (Tatschke et al. 2006). The early fusion approach is directly fusing the sensor's raw data. In the multi-level fusion approach, the perceived objects of all sensors are individually tracked in a sensor data fusion. Afterward, they are combined by a single fusion system. The grid-based approach processes the measurements by a grid-based fusion system. Consecutively, the grid data is used to generate object hypotheses. The track-level fusion approach focuses on fusing already built object hypotheses. The last approach, the fusion feedback approach, focuses on the possibility to feedback build object hypotheses into the fusion process to improve creation of object hypotheses.

5.2.4 Basic System Architecture for Sensor Data Fusion of the Project PRORETA 1

In the project PRORETA 1, see (PRO 2002-2006), a basic system architecture for sensor data fusion of environmental perception sensors has been developed, as described in (Darms 2007) or (Darms and Winner 2005). This architecture can only process object hypotheses based fusion systems. It is separated into three layers: the sensor layer, the fusion layer, and the application layer. The sensor layer takes care of the processing of the measurement data, the search for features, and the object classification, which is directly based on the measurement data. The fusion layer contains a classical object hypotheses based sensor data fusion. Beyond that, the layer may contain object classification algorithms as well as algorithms to prioritize object hypotheses. The architecture is completed by the application layer. This layer contains a concrete driver assistance application.

The architecture shown in (Darms 2007) describes its structures from the run time view only. The interfaces between the sensor and the fusion layer are described by a sensor specific data format. Between the fusion and the application layer, a scene graph is used. It describes object hypotheses by using an object oriented structure. Because the project PRORETA specifically focused on longitudinal traffic, the class vehicle-rear has been used in the concrete implementation. Furthermore, the interfaces between the components within the fusion process are mainly discussed from a project specific view and not in a general way.

5.2.5 Generic Sensor Data Fusion of the Institute of Measurement, Control and Microtechnology at Universität Ulm, Germany

Members of the Institute of Measurement, Control and Microtechnology at Universität Ulm, Germany, have considerable experience in perception systems. In (Dietmayer et al. 2005), they described basic architectural principles for sensor data fusion systems. This paper focused especially on the object hypotheses based sensor data fusion, as implemented in the fusion architecture of Volkswagen research department.

The idea of a general sensor data fusion system has been followed up. In the project Generic Sensor Data Fusion, see (Munz 2011), an application independent sensor data fusion system has been developed. The author showed six necessary steps plus the processing application to process the sensor data. At first, the measurement data is captured. This step is followed by a signal preprocessing step and a feature extraction step. Afterward, the features are associated and fused with already known data. In addition, the data is processed by an object management and, finally, sent to the consuming application. The challenge of application independence is conquered by using the concepts of state estimation, as shown in (Kämpchen 2007), and existence estimation, as introduced by (Mählisch 2009), simultaneously. Therefore, an application can use its own thresholds when to consider an object hypothesis or ignore it. To support the change of individual sensors within a concrete fusion system, there is a sensor independent interface between the Steps 1–3 and Steps 4–6. This interface consists of a state vector annotated by its static attributes (e.g. detection probability).

5.2.6 General Data Fusion Architecture of the University of Rochester, NY, USA

In (Carvalho and Heinzelman 2003), the authors presented a general data fusion architecture. In their paper, they showed especially the static architectural view. They defined three different classes of fusion systems: low-level fusion systems, high-level fusion systems, and hybrid fusion systems. The low-level fusion systems process only minor preprocessed sensor data. If the incoming data is already preprocessed by low-level fusions systems, the fusion system is referred to as high-level fusion system. In the case of a combination of high-level and low-level fusion systems, the authors speak of hybrid fusion systems. According to the paper, the architecture can be used in various ways due to the highly abstract description ranging from the military field to the monitoring of medical data.

5.2.7 Discussion of the Described Articles and Relation to this Paper

In the past, some papers in the field of architectures for environmental perception were published. The older articles, like (Bedworth and O'Brien 1999), Boyd (1976), and (Dasarathy 1997), built the foundation for the later papers. Because of their high grade of abstraction, they cannot be used directly as an implementation template for a concrete project. The same can be said for the architecture described in (Carvalho and Heinzelman 2003). The LAAS-Architecture did not get that much attention because the JDL-Architecture had already taken place as reference architecture for sensor fusion systems. In the automotive field, the Ph.D. thesis of Darms (2007) received a lot of attention. But his work only covers the runtime architectural view. Furthermore, only the object hypotheses based sensor fusion style is supported. Other kinds of fusion systems (e.g. grid-based fusion systems) are not supported. The PF2-Architecture supports object hypotheses based fusion systems as well as grid based fusion systems. Because of its high abstraction level, the PF2-Architecture can be used more as a reference architecture rather than helping with creating a concrete system.

The work described in (Munz 2011) is moving in another direction than in the previous architectures. The author follows the idea to create a sensor data fusion that is feasible for all applications. However, the following sections describe an architecture to design a specialized sensor data fusion for an individual application based on its requirements.

5.3 Scope of the Architecture

The architecture, shown in the next sections, is split into two different parts. Mathematically one can show that the object hypotheses based and grid structure based sensor data fusion share the same basic elements. But due to performance reasons, they are separated into their own architectural components.

The object hypotheses based sensor data fusion architecture specifies a sensor data fusion for environmental perception in a road vehicle. It aims at mapping the measurements from one or more sensors to an object hypotheses based data structure. An object hypothesis denotes the assumption that there is an obstacle in the real world. It is described by a set of variables. In this paper, this set is called a state. The measured input data can also be object hypotheses or just simple numerical values.

The grid based sensor data fusion architecture maps one or more measurements of one or more sensors to a grid based data structure. It is not important whether the measurements have been preprocessed. The grid data structure is described as the representation of a real world area by combination of two-dimensional areas (cells) whose dimensions are defined by a mathematical algorithm (e.g. offset surface, radial surface). Every grid has its own discrete coordinate system with a "cell" as its smallest unit. The coordinate system is defined relatively to another coordinate system, e.g. world fixed coordinates or vehicle fixed coordinates.

High-level fusion systems combining decisions by different arbiters, as presented by Rosenblatt (1997), or evaluating situations, as presented by Freyer et al. (2007), are not covered by the architecture.

5.4 Software Architecture of a Vehicle Environmental Perception System

Generally, software architecture is defined by Bass et al. as:

> "The software architecture of a program or computing system is the structure or structures of the system, which comprises software elements, the externally visible properties of those elements, and the relations among them. (Bass et al. 2003, p. 21)"

First of all, the requirements for defining the basis of the software architecture are presented. Because the experimental vehicle is used in different projects, the architecture cannot be defined in terms of the requirements of the Stadtpilot project only. Therefore, this paper focuses on the requirements on a meta-project-level. According to various requirements of different projects, different sensors have to be used or new interfaces need to be implemented. This can even result in a system configuration which does not need a sensor data fusion at all. The replacement of particular algorithms or the test of new approaches is elementary in the domain of research and also in vehicle systems engineering projects.

Briefly, this results in four requirements (ordered as covered in the following text):

1. The usage in a project which has no need for a sensor data fusion must be guaranteed.
2. The usage of different sensor configurations with different interfaces must be guaranteed.
3. The independence of individual processing stages and individual algorithms must be guaranteed.
4. The usage in various projects should be possible with low effort.

To meet these requirements, this software architecture is separated in multiple layers, as is the JDL-architecture presented by Steinberg et al. (1999). Figure 5.2 shows the layers of the Stadtpilot's architecture as UML-packages. The sensor specific layer conforms to Level 0 of the JDL-architecture and contains the processing of sensor specific protocols and bus systems as well as a possible preprocessing. The second layer corresponds to Level 1 and will be referred to as sensor data fusion layer. Its task is to process the data produced by the sensor specific layer and present the result to the application layer. The last layer is the application layer. It summarizes Level 2, 3 and 4 of the JDL-architecture and contains the specific function of the vehicle, e.g. an automatic emergency brake to reduce the consequences of an accident.

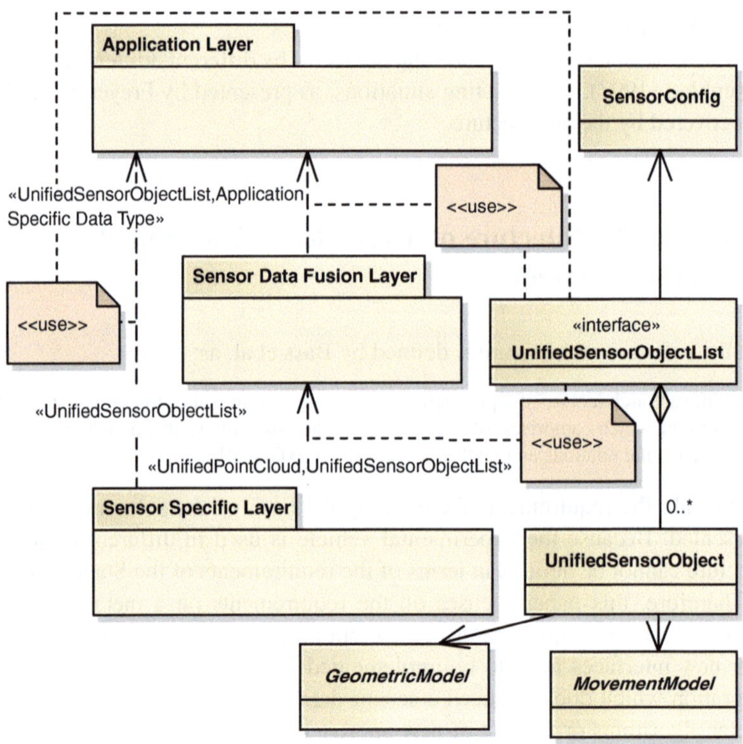

Fig. 5.2 Component diagram of the software architecture

In the case of the object hypotheses based sensor data fusion, the interfaces between the layers are designed to be compatible. This results in a system configuration in which an application can use the sensor's data directly and skip the sensor data fusion layer. Therefore, Requirement 1 is fulfilled. To design the interfaces as general and adaptable as possible, a container format for object hypotheses is defined. The data structure *UnifiedSensorObjectList* contains the mounting position, detection areas of the sensors, and the object hypotheses. Depending on the sensor, different models to represent the sensor's object hypothesis geometric and motion representation capabilities can be used. Sensor specific classifications and flags will not be transferred to enable uniform processing in the following layers.

For the grid based data structure, another container format has been defined due to performance reasons: *UnifiedPointCloud.* It contains measured point clouds in a 3D-Cartesian coordinate system as well as the sensor's mounting position. If necessary, the grid based sensor data fusion can also process *UnifiedSensorObjectLists* depending on the used sensor models.

The processing of a sensor specific protocol is represented by one component per each sensor type within the sensor specific layer. It possesses a sensor specific input to receive measurements and a sensor specific output to stimulate the sensor, as well as an output for the container formats. By these uniformed interfaces, sensors can be replaced easily, thus fulfilling Requirement 2.

The sensor data fusion layer may contain different object hypotheses or grid based sensor data fusions depending on the application specific requirements. Every sensor data fusion consists of a fusion module that is designed by the same scheme (see Sect. 5.6 and Sect. 5.8). Grid structure based sensor data fusion modules and object hypotheses based sensor data fusion modules may exchange data by using the defined container formats.

In the following sections, a tracked object hypothesis of a fusion module will be referred to as track. Referring to (Blackman and Popoli 1999, p. 595), an object hypotheses based sensor data fusion contains the components association and estimation. Additionally, a database for permanent storage of tracks is needed. These basic elements will be complemented by a time triggered component for track management and a basic component providing, e.g., static data for evaluation purposes and taking care of the overall fusion process. The individual interfaces between these components will be discussed in Sect. 5.6. They implement Requirement 3, which postulates the considerable independence of the used algorithms.

In the grid structure case, one grid structure based sensor data fusion module may contain one or more grid structure based data structures. One individual grid data structure will be referred to as grid layer in the following sections. Referring to (Thrun et al. 2005, p. 288), the algorithm of a grid structure based sensor data fusion consists of an (inverse) sensor model, a filter and a data storage. These basic elements are supplemented by a presentation component (*View*) to the application layer. The individual interfaces between the components are discussed in Sect. 5.8. They also implement Requirement 3.

The application layer processes data created by the sensor data fusion layer or the sensor specific layer by receiving *Unified Sensor ObjectLists* or *Unified Point Clouds*.

In this layer, the vehicle's actual function is implemented. Because of the two preceding layers, the function is largely isolated from the sensors. Therefore, the usage of a sensor data fusion in various projects is simplified significantly (Requirement 4).

5.5 Sensor Specific Layer

The data of every type of sensor is processed by a class adapted to this special sensor (e.g. *IBEOAlaskaSensorDeEncoder*). This class is derived from the abstract basic class *SensorDataObjectDeEncoder* and/or the abstract basic class *SensorDataPointCloudDeEncoder*. It provides service functions (e.g. coordinate transformations) as well as a frame for processing the sensor's data. The reception of measurement data is implemented by the class *SensorDataDecoder*. Figure 5.3 shows a class diagram of this component. The concrete sensor specific class performs two main actions: the de- and encoding of the sensor specific protocol and sensor specific tasks as the sensor's stimulation with the vehicle's ego motion data. The above mentioned container forwards *UnifiedSensorObjectList* and *UnifiedPointCloud* from the interface to the next layers.

In the *UnifiedSensorObjectList* structure, the data of the individual object hypotheses are stored. Depending on the sensor's capabilities, different geometric descriptions (e.g. point or box object hypothesis) as well as different motion descriptions (e.g. constant motion, motion with constant acceleration or no motion information) can be used. The *UnifiedPointCloud* data structure is used to transmit 3D-point clouds measured by the sensor. Figure 5.3 shows an example implementation of two different sensors with similar protocols sharing a common protocol decoding part. They both derive from the *IBEOSensorDeEncoder* which inherits from both abstract basis classes. Therefore, both container formats are provided by each of the child classes.

5.6 Object Hypotheses Based Sensor Fusion

Depending on the application requirements, the sensor fusion layer may contain several object hypotheses based sensor data fusions named object fusion modules. Based on the requirements mentioned above, algorithms ranging from very costly to very simple ones can be used within such a module. In the following, the coarse processing within a fusion module will be presented followed by the internal structures of the contained components except the *TrackDatabase* component.

The coarse cycle within a fusion module has already been described in Sect. 5.4. Figure 5.4 shows the components of an object fusion module (described in the next sections). It is built on top of the pipes-and-filter-pattern as described by Buschmann et al. (1996, p. 20). Every processing stage can work in parallel to the others and communicates the results of its algorithms by interfaces. This principle enables the reuse of individual algorithms in other fusion modules because of the separation of algorithms (Requirement 3). Therefore, an implemented and tested measurement association component can be used within a fusion module with an experimental filter component.

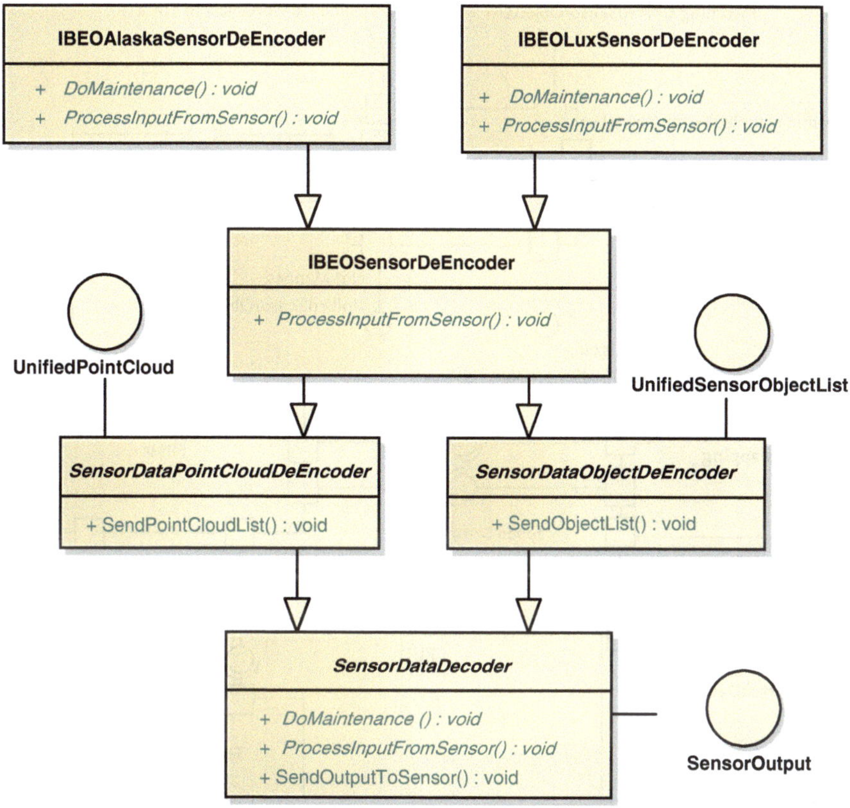

Fig. 5.3 Class diagram of the sensor specific layer

Darms refers to two different classes of sensor data fusions, which can be combined. In his Ph.D. thesis (Darms 2007, p. 25), he differentiates between central and distributed, as well as synchronous and asynchronous sensor data fusions. Because every object fusion module within the sensor fusion layer has the same input and output interfaces, they can be stacked. Consequentially, distributed as well as central sensor data fusion structures can be built. To create a synchronous sensor data fusion the input data can be queued and processed in a conjoint fusion step. Figure 5.5 shows an exemplary fusion module based on this architecture containing central, distributed, synchronous, and asynchronous fusion modules.

5.6.1 Basic Component

The *Basic* component of an object fusion module provides global control on the fusion cycle as well as starting and stopping the object fusion module. Moreover, the

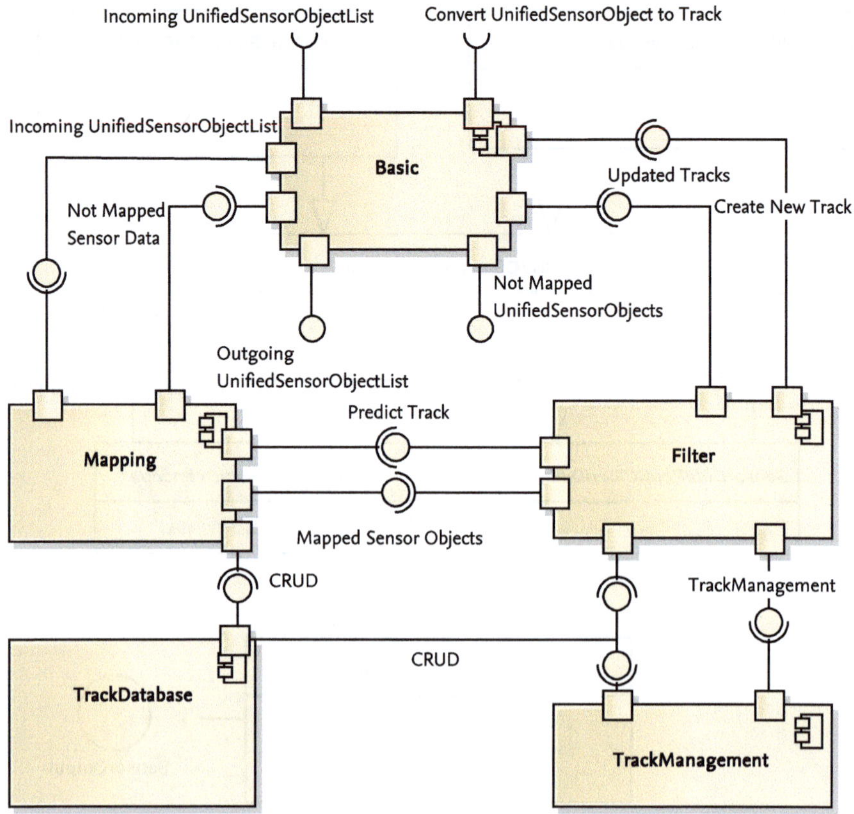

Fig. 5.4 Component diagram of an object fusion module within the sensor fusion layer

input and output interfaces to the other layers are implemented in this component. Besides, the interface for incoming and outgoing *UnifiedSensorObjectLists*, i.e., an interface for an object hypothesis which cannot be mapped to an already known track is implemented. This object hypotheses interface can be connected to an incoming interface which creates a new track from every incoming object hypothesis. By monitoring this data stream, the creation of new tracks can be prevented and the creation of tracks from false targets or from implausible data can be reduced.

5.6.2 Mapping Component

The *Mapping* component is responsible for assigning a measured object hypothesis of a sensor to a track already known by an object fusion module. As depicted in Fig. 5.6, two individual classes need to be implemented: *MappingAlgorithm* and *MappingMetric*. The *MappingAlgorithm* implements a mapping of object hypothe-

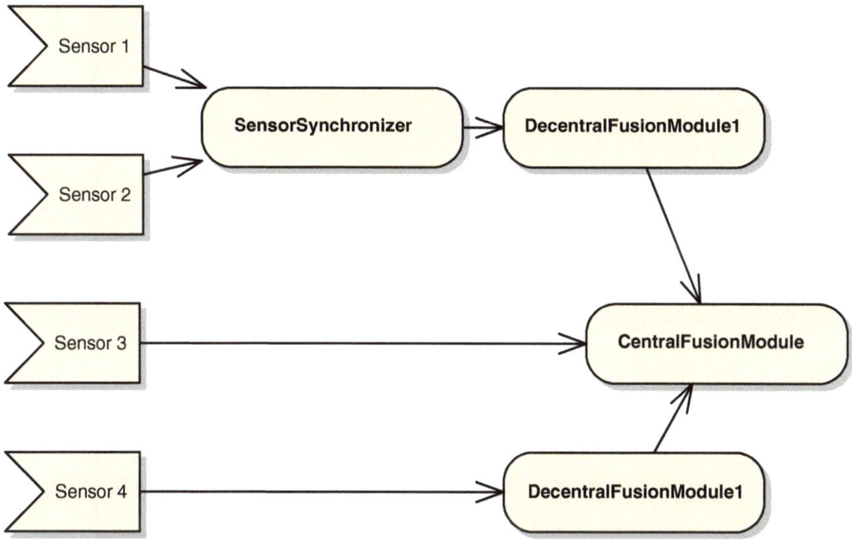

Fig. 5.5 Hybrid object fusion module containing central, distributed, synchronous and asynchronous object fusion modules

ses to tracks already existing in the object fusion module [e.g. Auction algorithm (Bertsekas 1979) or Hungarian algorithm (Munkres 1957)]. This results in one or more 3-tupels per object hypothesis containing the object hypothesis, the track, and the quality of the mapping. If it is not possible to map an object hypothesis to any track, it is sent to the interface for not mapped object hypotheses provided by the basic component. Every mapping algorithm uses a metric [e.g. Euclidian distance (Deza and Deza 2009, p.104) or Mahalanobis distance (Mahalanobis 1936)]. This class *MappingMetric* depends on the used geometric and motion model of the object hypothesis and the track. It calculates a quality for mapping a specific object hypothesis to a specific track.

By splitting the mapping component into two individual classes, these classes can be used in different combinations. Therefore, new algorithms can be realized by using already implemented metrics or fusion modules from other projects can be altered to use another algorithm (Requirements 3 and 4)

5.6.3 Filter Component

The *Filter* component of an object fusion module is connected to all other components. It forms the interface between the *Mapping* component and the application. Furthermore, it represents a part of the *Track Management*. The object hypothesis model used for the application interface is selected here as well. Within a *Filter*

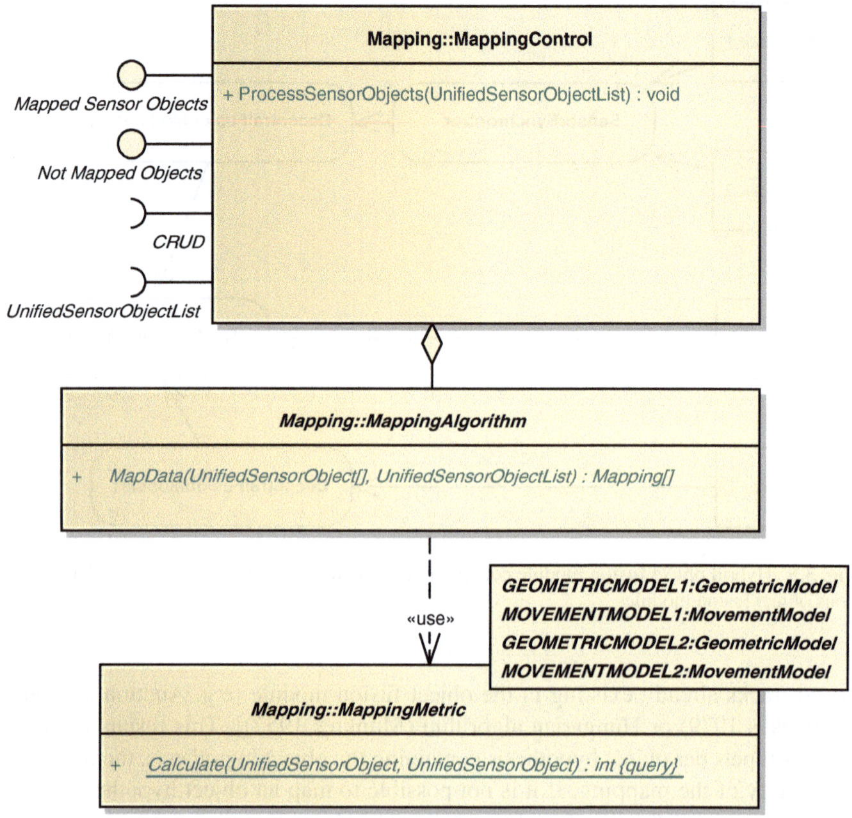

Fig. 5.6 Class diagram of the *mapping* component; CRUD: Create, Read, Update, Delete

component, different filter types can be implemented, e.g. Kalman filter (Kalman 1960), Particle filter (Eidehall et al. 2005) or Interacting Multiple Model filter (Blom 1984). To process different object hypotheses from various sensors, special functions are needed to handle measured states within the filtering process. A *Filter* component consists of two classes: *FilterAlgorithm* and *FilterUpdate* (Fig. 5.7). The *FilterAlgorithm* consists of the following individual functions: prediction of a track, update of a track, creation of a track, and filter specific track management. The class *FilterUpdate* is used to handle different object hypothesis models during a track update. It offers a function to handle a specific object hypothesis model within a concrete filter type as well as a function to convert an object hypothesis model to the one used by the *Filter* component.

Creation of a new track is initiated by the *Basic* component. It calls the function for track creation of a *FilterAlgorithm* class with an object hypothesis. This function creates the filter specific data structures. Afterward, the class *FilterUpdate* is called to convert an object hypothesis model to the filter's model. The result is saved in the

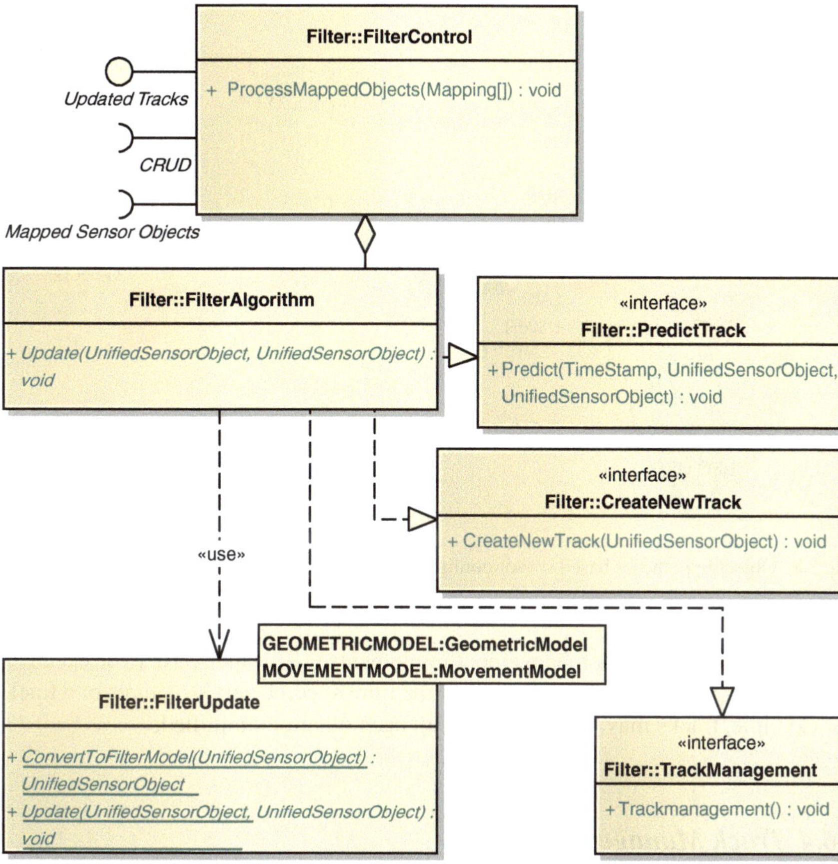

Fig. 5.7 Class diagram of the *filter* component ; CRUD: Create, Read, Update, Delete

TrackDatabase via the CRUD-Interface (Create, Read, Update, Delete) and can be mapped to new incoming object hypotheses in future fusion cycles.

The *FilterUpdate* combines a measured object hypothesis with a track. As an input data, the 3-tupel calculated by the mapping component is used. In case of the widely used Kalman filter, an update process is structured as follows: At first, a prediction of the track to the time of the measurement is performed. It is followed by the method *FilterUpdate:Update* which creates a measurement matrix. The process is finalized by calculating the Kalman filter's update step. The result is stored in the *TrackDatabase* again.

The prediction of a track to another point in time is very filter specific and, therefore, cannot be discussed in this paper in detail. The algorithm has to create a copy of the track and put the predicted data in this copy.

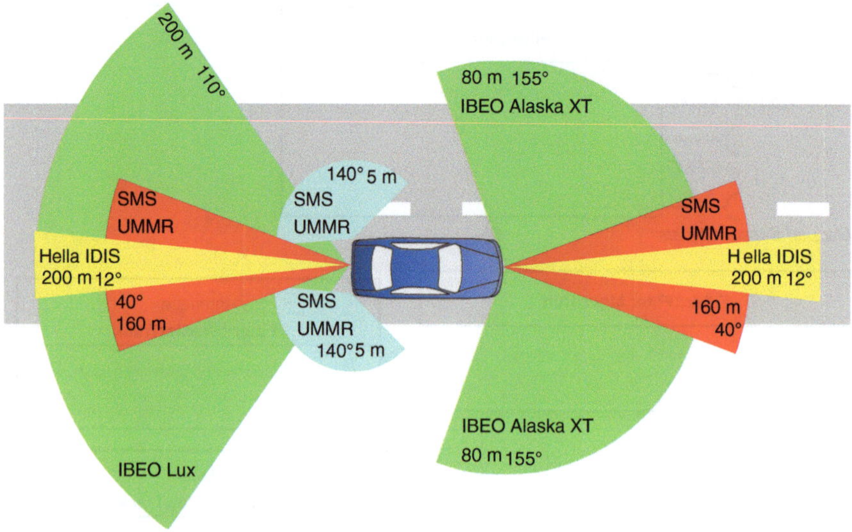

Fig. 5.8 Object hypotheses based sensor configuration of the Stadtpilot's test vehicle "Leonie"

The filter specific track management implements tasks which have to be executed on a time triggered basis and take care of the filter specific data structures of a track. For example, tracks may be combined with each other or simplified.

5.6.4 Track Management

The main task of the *TrackManagement* is to reduce tracks which are no longer needed and, therefore, will not receive any more updates. The time stamp of the last change and the time of creation are saved to the *TrackDatabase*. That way, this task is simplified. Furthermore, the filter specific track management function is also called to perform complex filter specific maintenance tasks.

5.7 Operational Application of the Object Hypotheses Based Sensor Fusion

The architecture elements presented in the preceding sections have been implemented for different use cases in the Stadtpilot project. For assembling the project's fusion modules, two different mapping algorithms can be used. A simple local-nearest-neighbor, as discussed by blackman and Popoli (1999, p. 9), as well as a minimum filter, as described by Schneider (2006, p. 56), realizing a simple mapping algorithm on the basis of a matrix were implemented. These algorithms can be used with a metric based on the Euclidian distance (Deza and Deza 2009, p. 104) between measurement and track. Furthermore, a variation of this metric with adaptive thresholds was implemented. Currently, three filter components can be selected: A Kalman filter with x-y-position object hypothesis model and motion vectors in x- and y-direction, a contour estimating Kalman filter, as described by Ohl and Maurer (2011), and a nonoperational filter for testing purposes which simply copies the measurement data to the track. Depending on the application requirements, up to twelve different fusion modules can be created from these parts.

The experimental vehicle of the Stadtpilot project is currently equipped with six different types of sensors used by the object hypotheses based sensor fusion (see Fig. 5.8). At the vehicles front bumper two IBEO Alaska XTs and a Hella IDIS 2 LIDAR scanner are installed alongside a Hella IDIS static LIDAR, and a SMS 2010 midrange RADAR system. The vehicle's rear bumper is equipped with one IBEO Alaska Lux LIDAR scanner, another Hella IDIS static LIDAR, and three SMS 2006 RADAR systems.

All sensors can be processed by the main sensor data fusion module of the project. For this object fusion module, the minimum filter is used with Euclidian distance metric combined with the contour estimating Kalman filter, which describes object hypotheses by an open polygon. This fusion module is accompanied by a fusion module for determining of the stability of a track by tracking special points on the hypothesis contour. In this second fusion module, a simple local-nearest-neighbor

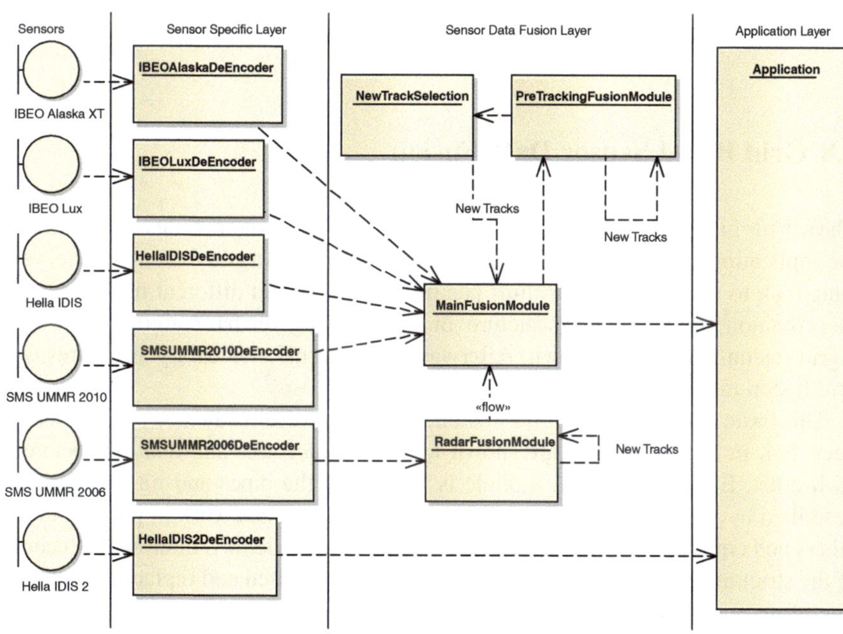

Fig. 5.9 Implementation of the architecture for the Stadtpilot project

mapping component with the adaptive Euclidian distance gating is combined with the one point Kalman filter. For the estimation of lateral velocities from RADAR object hypotheses, a fusion module with local-nearest-neighbor algorithm for measurement mapping with a simple Euclidian distance metric and the one-point Kalman filter was implemented.

Figure 5.9 shows the assembled modules as they are connected to each other. On the left side, one can see the different sensor types whose data is decoded by *Sensor-DeEncoders* for each sensor. Afterward, all measured object hypotheses are available as *UnifiedSensorObjectList* and can be sent to the *MainFusionModule*. The RADAR data has been processed by the *RadarFusionModule* before. The *PreTrackingFusion-Module* represents the fusion module for track stability analysis. Tracks successfully tracked by this module are sent to the *MainFusionModule* for track creation. The data stream is observed by a module for object initialization (*NewTrackSelection*) which selects object hypotheses based on the sensor's detection areas and denies them becoming a track to reduce false targets. Describing this algorithm in detail is outside of the scope of this paper.

By the availability of existing algorithms and the interface definitions within an object fusion module, enhancements and new algorithms can be easily added to already existing object fusion systems (Requirement 3). Because of unified interfaces between the layers, the Requirement 1 for directly accessing sensor data in an application has been fulfilled. The usage of an object fusion module in various projects results also from the unified interfaces between layers (Requirement 1). The usage of different sensor configurations (Requirement 2) has been fulfilled by the unification of the interfaces of the sensor protocol decoding.

5.8 Grid Based Sensor Data Fusion

The sensor fusion layer may contain one or more grid fusion modules depending on the application's requirements (see Ohl et al. 2011). Within a grid fusion module, very simple up to very costly algorithms can be used to convert different measurements to a common grid based data structure. In the following, the basic algorithms within a grid fusion module are shown. Afterward, the internal elementary structures of a grid fusion module are presented.

The basic elements of a grid fusion module have already been described in Sect. 5.4. In Fig. 5.10, they are shown with their interfaces and relations between each other. Every grid fusion module is built upon the pipes-and-filter-pattern, as described by Buschmann (1996, p. 200). Every element can work in parallel to the others and communicates its algorithm's results by well-defined interfaces. Because of the strict modularization, algorithms can be easily reused and replaced by others. Therefore, a production ready *SensorModel* could be used in combination with an experimental *Filter* (see Requirement 3).

Within a grid fusion module, components may be used several times. This enables the usage of grid layers with different resolutions or the usage of different sensor

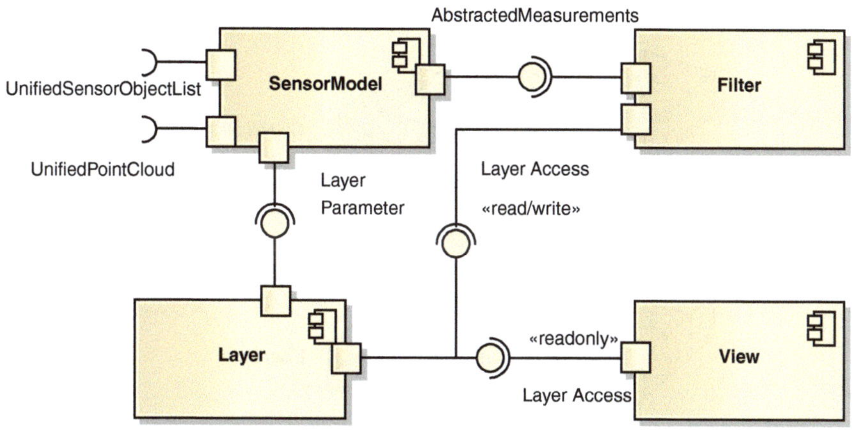

Fig. 5.10 Component diagram of a grid fusion module

characteristics (e.g. height resolution). Two grid fusion modules will be separated if they follow a different objective. If the same area around a vehicle is observed by two grid layers with the objective "obstacle avoidance", we speak of one grid fusion module. If one of the grid layers has the objective "obstacle avoidance" and the other has the objective "height mapping", we speak of two different grid fusion modules.

5.8.1 Layer Component

The *Layer* component is the central component within a grid sensor fusion module and represents a black-board-pattern (Buschmann et al. 1996, p. 205). It stores and manages the actual data. One individual component represents a single grid data structure. This grid structure may correlate with other grid structures. If these relations are projected on the third dimension, we speak of multiple *Layers* (logical grid structures). During operation, a decentralized data fusion, as described by Darms (2007, p. 25), can be constructed with these layers. E.g., the measurements of every sensor are processed in an individual *Layer* instance, after the first processing stage these data can be fused in a combination *Layer*. Figure 5.11 shows this example. In this case, the measurements of a RADAR sensor and a LIDAR sensor are processed in an individual *Layer* and then combined in a shared one.

In Fig. 5.12 the class diagram of the *Layer* component is depicted. The class *Layer* has multiple attributes to manage the grid data structure. As shown above in Sect. 5.3, a *Layer* has its own coordinate system supplemented by the dimension of a cell and the count of cells. To allow parallel operations on as many cells as possible, the data structure has been separated into multiple *GridBlocks*. Shifting and extending a grid data structure is also made easier by *GridBlocks* (Fig. 5.13). A *GridBlock* represents a micro-grid and takes care of actually storing the data. Every *GridBlock* can be

Fig. 5.11 Logic grids (Layer), e.g. RADAR Layer, LIDAR layer und combination layer

locked for read and write operations individually. Therefore, two *Filter* components can work on the grid data structure in parallel if their detection areas do not intersect. The interfaces of the *Layer* class are used by the other components to interoperate with the class and, e.g., to transform coordinates or to update a cell's content.

5.8.2 Sensor Model Component

A sensor model component implements the mapping of measurements to the values needed by the *Filter* to update a single cell, as stated by Thrun er al. (2005, p. 288). There could be a special model for every type of sensor. Above all, differences between *SensorModels* can be created on the basis of the sensor technology. It is obvious that in the case of a LIDAR sensor model there are other cells to update than in the case of a sensor model built for RADAR measurements.

The abstract class *SensorModel* with its interfaces is shown in Fig. 5.14. The interface to the *Filter* component is represented by a list of cells to update and a special filter specific data structure. The figure shows example structures for a Dempster-Shafer filter, as introduced by Shafer (1976), and a Bayes filter, as described by Park (2007, p. 45). Because the filter should process complete lists only, the update list is only sent once for every *UnifiedPointCloud* processed by the *SensorModel* component. This results in a consistent data structure in the *Layer* component.

As the *UnifiedPointCloud* data structure contains measurement points only, the information of the used sensor coordinate system has to be represented by the sensor model to update the correct cells within a layer. For a rotating LIDAR sensor with a radial coordinate system, whose measurements should be inserted into a *Layer* with a Bayes filter, the insertion process could be as follows: The *UnifiedPointCloud*

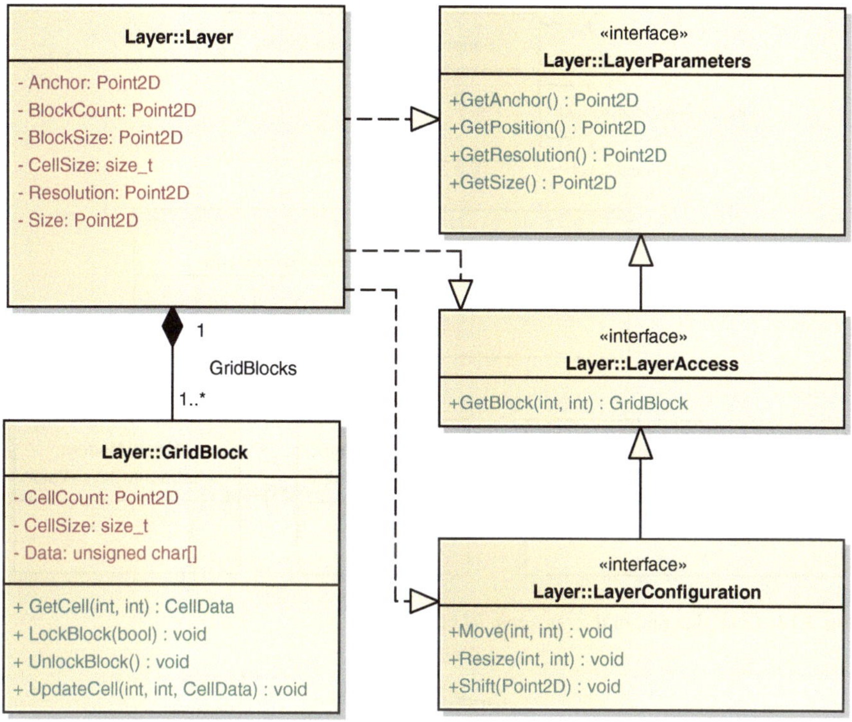

Fig. 5.12 Class diagram of the *layer* component

Fig. 5.13 Memory layout of a *Layer* containing multiple *gridblocks*

contains every laser measurement of the sensor. Objects that are located outside the detection range of the sensor are not represented. This results in "holes" in the point cloud and, therefore, has to be handled by the *SensorModel* component. If a *Layer* is updated by a measurement from an *UnifiedPointCloud*, the ray path beginning in the center of the sensor's coordinate system up to the measured cell is marked as non-occupied and the measured cell itself is marked as occupied (see Fig. 5.15, Object 1). Rays without a reflection within the sensor's range have to be estimated by the sensor model. Based on the angle resolution of the sensor, the measurements within the *UnifiedPointCloud's* data can be analyzed to handle these "holes". These

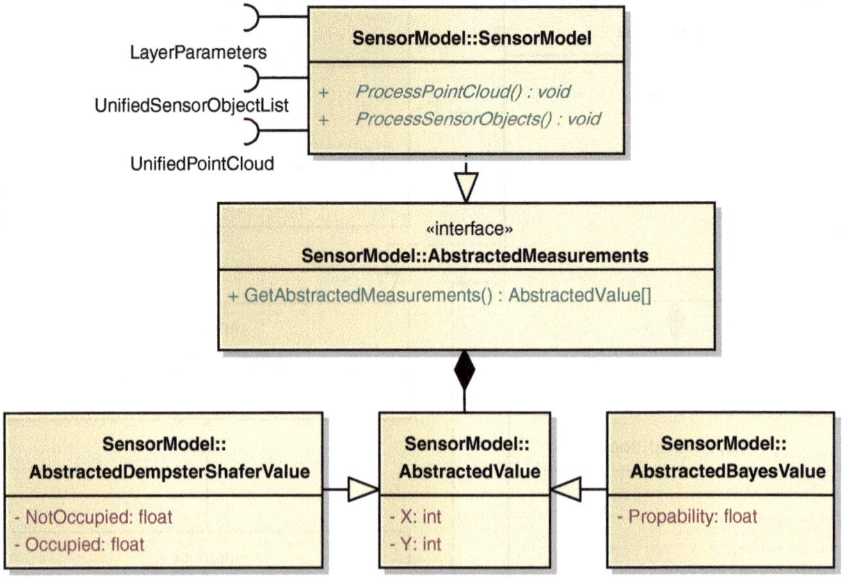

Fig. 5.14 Class diagram of the *sensormodel* component

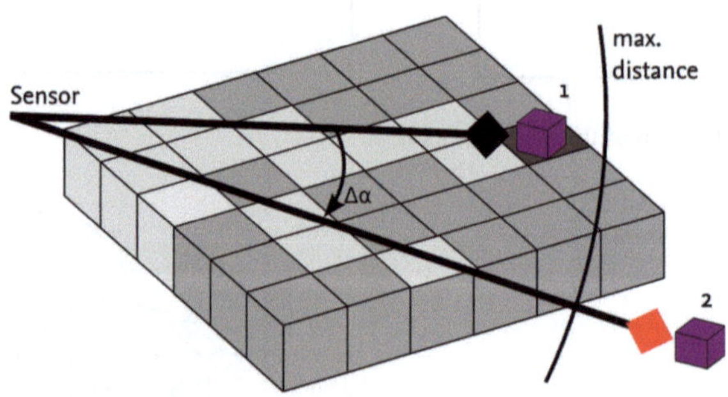

Fig. 5.15 Example of the processing of a rotating LIDAR sensor in a sensor model calculating an occupied (dark), non-occupied (light) property, object 1 within, object 2 outside the sensor's detection range

data can be used to update the rays along these non-existing measurements up to the sensor's maximal detection range as non-occupied (see Fig. 5.15, Object 2).

5.8.3 Filter Component

The *Filter* component updates the cells within a concrete *Layer* by using the results of the *SensorModel* component. To update a *Layer's* cells the Dempster-Shafer

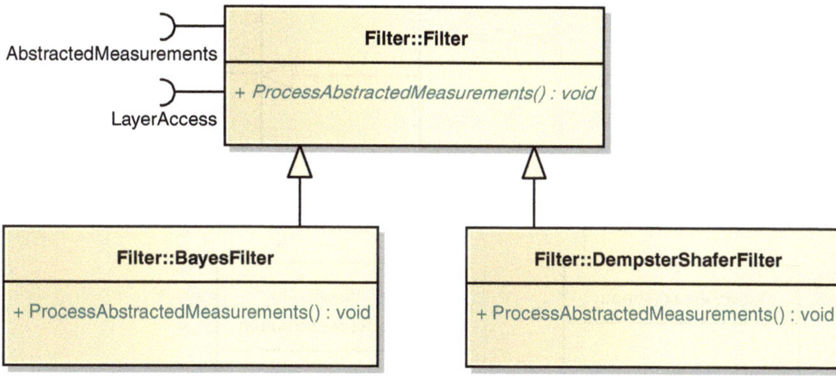

Fig. 5.16 Class diagram of the *filter* component

algorithm or the Bayes algorithm can be used, as well as several other methods. The actual data structure stored in a *Layer* results directly from the used *Filter* component and its implemented algorithm. During the combination of the abstracted measurements with concrete cells, the order of the measurements sent to the interface *AbstractedMeasurements* is important. If this list is ordered by grid blocks, the *Filter* component can save a lot of locking overhead and do the processing in parallel.

Figure 5.16 shows the class structure of the *Filter* component. It illustrates the interfaces used to access the *Layer* as well as to receive the results from a *SensorModel* component.

5.8.4 View Component

Aggregated data of the *Layer* component often have to be converted or preprocessed to be handled by the application layer. One example is to detect drivable areas in front of a test vehicle, as described by Leonard et al. (2008). To transmit the grid data structure to the processing application is not always possible and in most cases not a wise decision due to bandwidth considerations. The *View* component creates a special view of the grid data structure, especially for an individual application. By this means, very costly transformations which work with large parts of the grid structure can be performed directly on the *Layer's* data structures and do not have to be transmitted to an application.

The class diagram of the *View* component is depicted in Fig. 5.17. The basis class *View* is controlled by an external event (e.g. time triggered) and starts its algorithms based on this event. One algorithm implemented by a *View* component may be an algorithm to reconstruct objects from the grid structure. Other algorithms could copy the complete data to present it to the developer or to perform an occlusion area detection algorithm, see (Ulbrich 2011, p. 37).

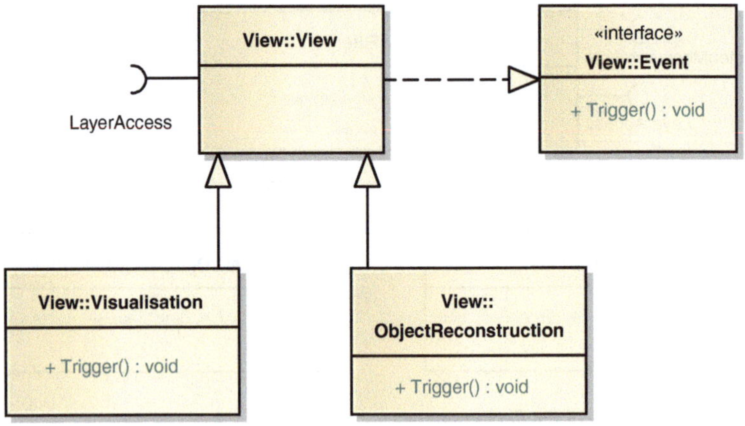

Fig. 5.17 *View* component's class diagram

5.9 Operational Application of the Grid Based Sensor Fusion

The components discussed in the previous sections have been implemented as part of the Stadtpilot project in the C++ programming language. The *SensorModel* component was realized for a scanning LIDAR sensor and can be parameterized to fit the date of different sensors. The *SensorModel's* result can be processed by a Bayes *Filter* component. The implemented layer data structure can be moved with the vehicle and also be used earth fixed.

The test vehicle of the Stadtpilot project is equipped with multiple sensor types. The grid based sensor fusion is currently only using the vehicle's LIDAR sensors (see Fig. 5.18). In the vehicle's front two scanning LIDAR sensors of type IBEO Alaska XT are mounted. The vehicle's rear is equipped with a scanning LIDAR of type IBEO Lux. At a central position on the vehicle's roof, a scanning LIDAR of type Velodyne HDL64ES2 is mounted.

Figure 5.19 shows the used sensors and the implemented modules and components with their relations. The used *Layer* component covers an area of 100 × 100 m and has a resolution of 20 × 20 mm. The *Layer* is separated into 8 × 8 *GridBlocks* à 64 × 64 cells. By now, the components process data of three sensors and combine them in a single *Layer* component. The implemented *SensorModel* component can be used for every sensor by using a different set of parameters. To process the measurements of the Velodyne HDL64ES2, a preprocessing is needed to remove measurements of the ground plane. The presentation of the grid's data is done by a component which completely copies the data structure and transmits it to a visualization application. The application uses the results of the grid structure based sensor data fusion for detection areas occluded by other objects, as shown by Ulbrich (2011, p. 37). An enhanced detection algorithm for lane width detection based on (Weiss 2011, p.128) has also been developed.

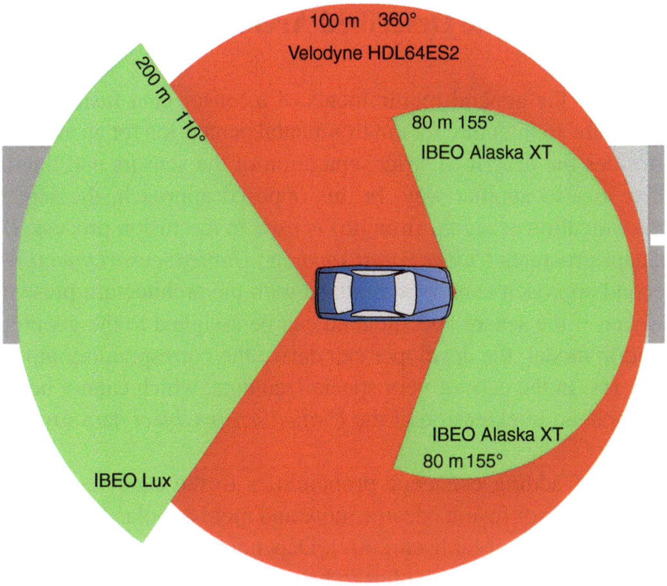

Fig. 5.18 Grid based sensor configuration of the Stadtpilot's test vehicle "Leonie"

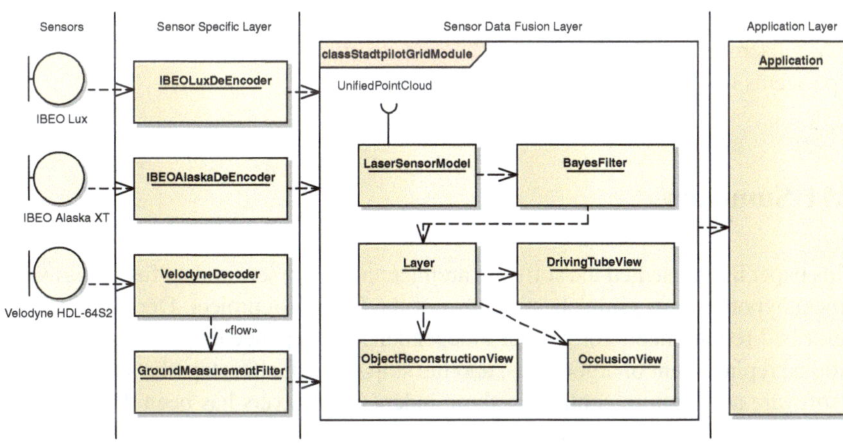

Fig. 5.19 Implementation of the architecture in the Stadtpilot project

By the availability of a great number of algorithms and the definition of interfaces within a grid fusion module, extensions to an existing or new fusion module can be done easily (see Requirement 3). The usage of independent *View* components enables the usage of a fusion module with applications of different projects (see Requirement 4). The Requirement 2 for using other sensor configurations has been realized by global interfaces to receive sensor data from the sensor specific layer.

5.10 Limitations of the Described Architecture

The realization of the general requirements of a sensor data fusion has led to the described software architecture for environmental perception for an automotive environment. Besides the described strict separation of the sensors and the data fusion process, there is also another way. In this opposed approach, the sensor specific data (e.g. classification or signal strength) is used in the fusion process. Because of the general data structures *UnifiedPointCloud* and *UnifiedSensorObjectList*, it seems that this second approach cannot be realized with the architecture presented in this paper. However, if the sensor specific data can be assigned to the geometric model or to the motion model, the developer can derive the corresponding model and add the needed fields. In the case of very specialized data, which cannot be assigned to either of the models, an extension of the *UnifiedSensorObject* data structure may be useful.

The concept of adding existence probabilities to the state vector of an object hypotheses based sensor fusion became more and more popular after the publication of Mählisch's phd-thesis (Mählisch 2009). Depending on the availability of a measurement's existence probability, it has to be transmitted by the sensor to the fusion algorithms. Because of the limitation of the *UnifiedSensorObjectList* data structure to two different models, this probability cannot be stored in this data structure and, therefore, not transferred to the application. To implement a fusion system according to Mählisch (2009), an extension of the data structure by a third probabilistic model type seems necessary.

5.11 Summary

This paper has presented the software architecture of the sensor data fusion based on object hypotheses or grid data structures of the Stadtpilot project. Derived from the described requirements for project independence, independence of sensor configurations, replacement of algorithms, and the skip of the fusion step, the static view of a software architecture containing three independent layers has been defined. This rough separation in sensor specific layer, sensor data fusion layer, and application has been described in detail; and the internal and external interfaces have been defined.

Altogether, the implementation of the fusion system of the Stadtpilot project has proved to be very straightforward. The further development of algorithms could be simplified while the reuse of components and algorithms in various projects and with different sensor configurations has been made possible by the exact definition of interfaces between and within the three layers.

References

Al-Dhaher, A., Mackesy, D.: Multi-sensor data fusion architecture. Proceedings of the 3rd IEEE International Workshop on Haptic, Audio and Visual Environments and Their Applications, HAVE, pp. 159–163. Ottawa (2004)

Bass, L., Clements, P., Kazman, R.: Software Architecture in Practice, 2nd edn. Addison-Wesley Professional, Boston (2003)

Bedworth, M.D., O'Brien, J.: The Omnibus model: a new model for data fusion?. Proceedings of the 2nd International Conference on Information Fusion, pp. 437–444. Fusion, Sunnyvale (1999)

Bertsekas, D.P.: A distributed algorithm for the assignment problem, Lab. for information and decision systems report. MIT, Cambridge (1979)

Blackman, S., Popoli, R.: Design and Analysis of Modern Tracking Systems. Artech House Publishers, Norwood (1999)

Blom, H.A.P., : An efficient filter for abruptly changing systems, Proceedings of the 23rd IEEE Conference on Decision and Control, pp. 656–658. CDC, Las Vegas (1984)

Boyd, R.R.: A Discourse on Winning and Losing: Slides Air University Library. Maxwell, AL (1976)

Buschmann, F., Meunier, R., Rohnert, H., Sommerlad, P., Stal, M.: Pattern-Oriented Software Architecture - A System of Patterns, vol. 1. Wiley, West Sussex (1996)

Carvalho, H.S., Heinzelman, W.B.: A general aata fusion architecture. Proceedings of the 6th International Conference on Information Fusion, pp. 1465–1472. Fusion, Cairns (2003)

Darms, M.: Eine Basis-Systemarchitektur zur Sensordatenfusion von Umfeldsensoren für Fahrerassistenzsysteme. Ph.D thesis, Technische Universität Darmstadt, Fachgebiet Fahrzeugtechnik (2007)

Darms, M., Winner, H.: A modular system architecture for sensor data processing of ADAS applications. In: Proceedings of IEEE Intelligent Vehicles Symposium IV Las vegas, pp. 729–734 (2005)

Dasarathy, B.: Sensor fusion potential exploitation-innovative architectures and illustrative applications. Proc. IEEE **85**(1), 24–38 (1997)

Deza, M.M., Deza, E.: Encyclopedia of Distances. Springer, Berlin (2009)

Dietmayer, K.; Kirchner, A. and Kämpchen, N.: Fusionsarchitekturen zur Umfeldwahrnehmung für zukünftige Fahrerassistenzsysteme. In: Maurer, M., Stiller, C. (eds.) Fahrerassistenzsysteme mit maschineller Wahrnehmung. Springer, pp. 59–88. Berlin Heidelberg (2005)

Durrant-Whyte, H.; Rao, B. Hu, H.: Toward a fully decentralized architecture for multi-sensor data fusion. Proceedings of IEEE International Conference on Robotics and Automation, Vol. 2 of ICRA, pp. 1331–1336. Cincinnat (1990)

Effertz, J.: Sensor architecture and data fusion for robotic perception in Urban environments at the 2007 DARPA Urban challenge. Proceedings of the 2nd International Conference on Robot Vision, pp. 275–290. RobVis, Auckland (2008)

Eidehall, A.; Schon, T., Gustafsson, F.: The marginalized particle filter for automotive tracking applications. In:Proceedings of IEEE Intelligent Vehicles Symposium IV Las vegas, pp. 370–375 (2005)

Freyer, J., Deml, B., Maurer, M., Färber, B.: ACC with enhanced situation awareness to reduce behavior adaptions in lane change situations. Proceedings of the IEEE Intelligent Vehicles Symposium, IV, Istanbul (2007)

Gad, A., Farooq, M.: Data fusion architecture for maritime surveillance. Proceedings of the 5th International Conference on Information Fusion, vol. 1 of Fusion, pp. 448–455. Annapolis (2002)

Kalman, R. E.: A new approach to linear filtering and prediction problems. J. Basic Eng. **82**(D), pp. 35–45 (1960)

Kämpchen, N.: Feature-level fusion of laser scanner and video data for advanced driver assistance systems. Ph.D thesis, Universität Ulm, Institut für Mess-, Regel- u. Mikrotechnik (2007)

Leonard, J., How, J., Teller, S., Berger, M., Campbell, S., Fiore, G., Fletcher, L., Frazzoli, E., Huang, A., Karaman, S., Koch, O., Kuwata, Y., Moore, D., Olson, E., Peters, S., Teo, J., Truax, R., Walter,

M., Barrett, D., Epstein, A., Maheloni, K., Moyer, K., Jones, T., Buckley, R., Antone, M., Galejs, R., Krishnamurthy, S., Williams, J.: A perception-driven autonomous Urban vehicle. J. Field Robotics **25**(9), 727–774 (2008)

Llinas, J., Bowman, C., Rogova, G., Steinberg, A., Waltz, E., White, F.: Revisiting the JDL data fusion model II. Proceedings of the 7th International Conference on Information Fusion, pp. 1218–1230. Fusion, Mountain View (2004)

Mahalanobis, P.C.: On the generalised distance in statistics. Proc. Natl. Inst. Sci. **2**(1), 49–55 (1936)

Markin, M., Harris, C., Bernhardt, M., Austin, J., Bedworth, M., Greenway, P., Johntson, R., Little, A., Lowe, D.: Technology Foresight on Data Fusion and Data Processing. The Royal Aeronautical Society, London (1997)

Mählisch, M.: Filtersynthese zur simultanen Minimierung von Existenz-, Assoziations- und Zustandsunsicherheiten in der Fahrzeugumfelderfassung mit heterogenen Sensordaten. Ph.D thesis, Universität Ulm, Institut für Mess-, Regel- u. Mikrotechnik (2009)

Munkres, J.: Algorithms for the assignment and transportation problems. J. Soc. Ind. App. Math. **5**(1), 32–38 (1957)

Munz, M.: Generisches Sensorfusionsframework zur gleichzeitigen Zustands- und Existenzschätzung für die Fahrzeugumfelderkennung. PhD thesis, Universität Ulm, Institut für Mess-, Regel- u. Mikrotechnik (2011)

Ohl, S., Matthaei, R., Müller, M., Maurer, M.: Softwarearchitektur der gitterbasierten Sensordatenfusion desProjekts Stadtpilot. In: Intelligente Transport- und Verkehrssysteme und -dienste Niedersachsen E.V. (eds.), AAET 2011, Automatisierungssysteme, pp. 281–297. Assistenzsysteme und eingebettete Systeme für Transportmittel, AAET, Braunschweig (2011)

Ohl, S., Maurer, M.: A contour classifying kalman filter based on evidence theory. Proceedings of the 14th International IEEE Annual Conference on Intelligent Transportation Systems, pp. 1392–1397. ITSC, Washington, DC, USA (2011)

Park, S.B.: ProFusion2 - D15.12 Final Report: Accessed online on 25.10.2011 http://prevent.ertico.webhouse.net/download/deliverables/ProFusion/202/PR-15000-SPD-v26/D15.12/ProFusion2/Final/Report.pdf (2007)

Polychronopoulos, A., Amditis, A.: Revisiting JDL model for automotive safety applications: the PF2 functional model. Proceedings of the 9th International Conference on Information Fusion, pp. 1–7. Fusion, Florence (2006)

PRO : PRORETA 1 - PRORETA: Accessed online on 26.04.2012 http://www.proreta.tu-darmstadt.de/proreta_1/projekt_proreta_1/index.de.jsp (2002–2006)

Rosenblatt 1997 Rosenblatt, J.: DAMN: A distributed architecture for mobile navigation. Ph.D thesis, Carnegie Mellon University, Robotics institute, Pittsburgh (1997)

Royce, W.W.: Managing the development of large software systems. Proceedings of the 9th International Conference on Software Engineering, pp. 328–338. ICSE, Monterey (1987)

Saust, F., Bley, O., Kutzner, R., Wille, J.M., Friedrich, B., Maurer, M.: Exploitability of vehicle related sensor data in cooperative systems. Proceedings of the 13th International IEEE Conference on Intelligent Transportation Systems, pp. 1724–1729. ITSC, IEEE (2010)

Saust, F., Wille, J.M., Lichte, B., Maurer, M.: Autonomous vehicle guidance on braunschweig's inner ring road within the Stadtpilot project. Proceedings of the Intelligent Vehicles Symposium, IV, pp. 169–174. Baden-Baden (2011)

Scheunert, U., Lindner, P., Cramer, H., Tatschke, T., Polychronopoulos, A., Wanielik, G.: Multi level processing methodology for automotive applications. Proceedings of the IEEE Intelligent Transportation Systems Conference, , pp. 1322–1327. ITSC, Toronto (2006)

Schneider, U.: Sensordatenfusion und Fehlerkalibrierung von umfelderkennenden Sensoren eines Straßenfahrzeuges. Ph.D thesis, Technische Universität Braunschweig, Institut für Regelungstechnik (2006)

Shafer, G.: A Mathematical Theory of Evidence. Princeton University Press, Princeton (1976)

Steinberg, A.N., Bowman, C.L., White, F.E.: Revisions to the JDL data fusion model. Sensor Fusion: Archit. Algorithms Appl. III **3719**(1), 430–441 (1999)

Tatschke, T., Park, S.-B., Amditis, A., Polychronopoulos, A., Scheunert, U., Aycard, O.: ProFusion2 - towards a modular, robust and reliable fusion architecture for automotive environment perception. In: Valldorf, J., Gessner, W. (eds.) Advanced Microsystems for Automotive Applications, pp. 451–469. Springer, Berlin Heidelberg (2006)

Thrun, S., Burgard, W., Fox, D.: Probabilistic Robotics. MIT Press, Cambridge (2005)

Ulbrich, S.: Intelligent decision making and maneuver planning for autonomous sriving in urban traffic environments. Diplomarbeit,Technische Universität Braunschweig, Institut für Regelungstechnik (2011)

Weiss, T.T.: Hochgenaue Positionierung und Kartographie mit Laserscannern für Fahrerassistenzsysteme. Ph.D thesis, Universität Ulm (2011)

White, F.E.: A model for data fusion. Proceedings of the 1st National Symposium on Sensor Fusion, vol. 2. Orlando (1988)

Chapter 6
Maneuver-Based Vehicle Guidance Based on the Conduct-by-Wire Principle

Sebastian Geyer

6.1 Introduction

6.1.1 Motivation

Modern Advanced Driver Assistance Systems (ADAS) have established a standard of driving comfort and safety unknown so far. Vehicles have become increasingly "intelligent" allowing the driver to delegate specific subtasks of vehicle guidance to these systems or to let the automation take over vehicle guidance completely in emergency situations. However, the scientifically proven advantages of ADAS (Hummel et al. 2011) are accompanied by an important and relevant side effect: increasing complexity. Today, most ADAS are developed separately, with the consequence that each of these systems has its own user interface and interaction concept. This increasing complexity is contrary to the original goal of enhancing comfort and safety.

The highest level of vehicle guidance automation is represented by fully autonomous cars. Research in this field has made extensive advances in the last ten years, but the challenges of fully autonomous vehicles in public traffic are just starting to be identified (Urmson et al. 2008; Saust et al. 2011; Ardelt et al. 2012). In particular, the question of possible approval processes for autonomous vehicles in public traffic is still unsolved. Moreover, today's legal requirements such as the 1968 Vienna Convention on Road Traffic (United Nations 1968) prevent the market introduction of these systems.

A solution for the described problem of increasing user complexity when combining multiple assistance systems and a very important step towards fully automated driving might be the innovative vehicle guidance frameworks of cooperative guidance and control, exemplified in concepts like H-Mode (Flemisch et al. 2003) or

S. Geyer (✉)
Institute of Automotive Engineering, Technische Universität Darmstadt,
Petersenstraße 30, D-64287 Darmstadt, Germany
e-mail: geyer@fzd.tu-darmstadt.de

M. Maurer and H. Winner (eds.), *Automotive Systems Engineering*,
DOI: 10.1007/978-3-642-36455-6_6, © Springer-Verlag Berlin Heidelberg 2013

Conduct-by-Wire (CbW). The driver of a Conduct-by-Wire vehicle assigns maneuver commands via the so-called maneuver interface, which also allows the parameterization of the chosen maneuvers and interaction at the stabilization level if desired. This concept allows a high degree of automation, while—unlike fully automated concepts—still keeping the driver responsible for the vehicle guidance according to the 1968 Vienna Convention on Road Traffic. The interdisciplinary research project Conduct-by-Wire at the Technische Universität Darmstadt aims to provide a technical feasibility assessment of the Conduct-by-Wire concept.

6.1.2 Objectives

This chapter proposes a methodology for the technical feasibility assessment of an innovative vehicle guidance concept based on the Conduct-by-Wire principle. Focus lies on the analysis of the driver-vehicle interaction. The methodology of the proposed top-down approach that allows an assessment in the early concept phase is depicted in Fig. 6.1.

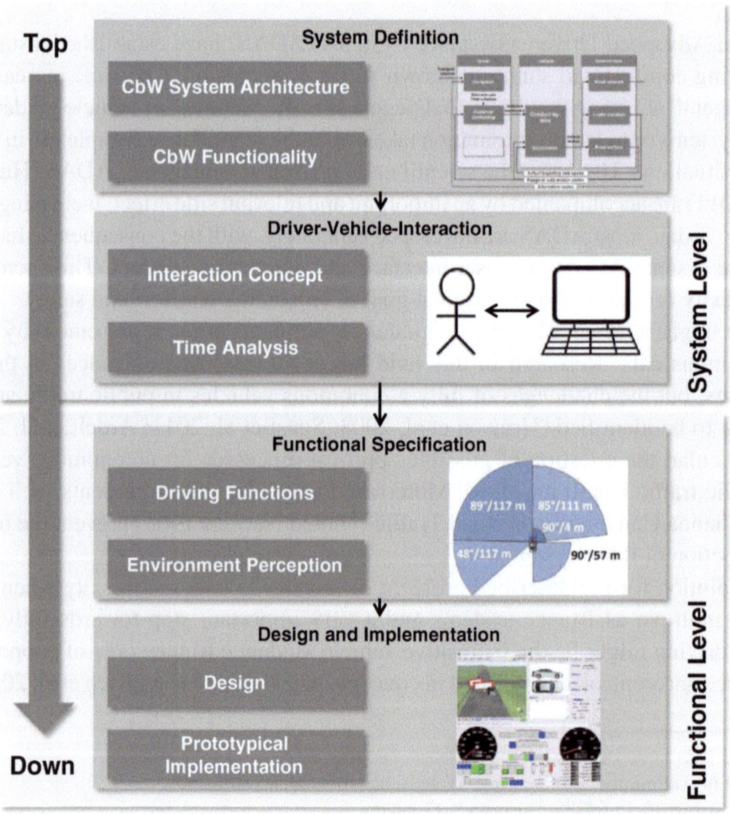

Fig. 6.1 Methodology for the technical feasibility assessment of Conduct-by-Wire (CbW)

Starting at the system level with the development of the CbW system architecture, the CbW functionality and a cooperative interaction concept—the gate concept—are concretized stepwise. Moreover, the driving situations this cooperative vehicle guidance concept has to cope with are systematically derived. These steps build the basis for analyzing whether the gate concept, as a theoretical approach to cooperatively pass the decision points along the planned trajectory during the maneuver execution, might also be suitable in practical use. The assessment is focused on analyzing the time available to the driver or the automation for the decision-making process.

These studies at the system level build the basis for concretizing the functional development, which in the following is exemplarily demonstrated for determining the requirements of the environment perception system.

Following this systems engineering approach makes it possible to handle the complexity and obtain a stepwise development. Furthermore, a decision on whether the realization of this innovative concept is worthwhile can be made at different abstraction levels before a prototype system has to be built.

6.2 Methodology

The technical feasibility assessment of CbW is mainly based on three pillars. The first is the development of the CbW system architecture that helps to concretize the system limits, the CbW functionality, and—the focus of this chapter—the driver-vehicle interaction. The second pillar is formed by requirements this vehicle guidance concept has to fulfill in terms of the driving scenarios that the system (driver, driver and automation, or automation only) has to cope with, which are derived from a driving situations catalog. Finally, the simulation environment IPG CarMaker is used to evaluate first concepts and to develop the automation's functionality.

6.2.1 Conduct-by-Wire System Architecture

In the past decades of vehicle automation development many different system architectures have been realized. The different approaches are described in detail by Lotz (Lotz 2013).

In the Conduct-by-Wire research project, adapting the three-level-hierarchy of the driving task by Donges (1982) has proved itself for both the driver-oriented and the automation-oriented concretization of Conduct-by-Wire in this early concept phase.

The three-level-hierarchy of the driving task by Donges illustrates the different levels at which the interaction between driver, vehicle and environment occurs. According to this model, vehicle guidance can be represented by three cascaded control loops. On the highest level, the navigation level, the route and possible modifications are planned.

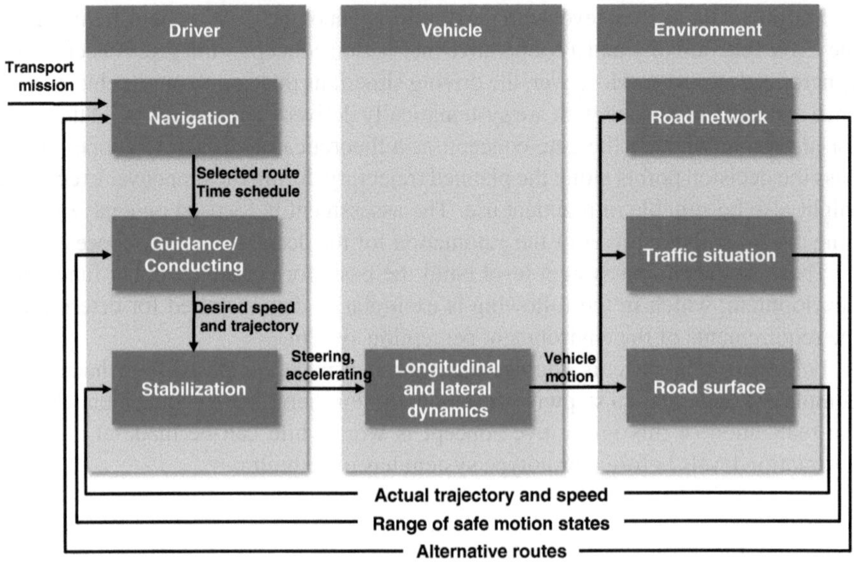

Fig. 6.2 Conventional vehicle guidance according to the three-level-hierarchy of the driving task by Donges (1982)

On the next lower level, the guidance level, the realization of the route is planned by means of driving maneuvers and trajectories—the latter implies the decision about the desired operation point in terms of speed, distance, lane, and position in lane—determined depending on the current situation. These output the set values for the stabilization task, which consists of minimizing the difference between actual and planned trajectory. Today, the dominant part of vehicle guidance is done at this stabilization level as shown in Fig. 6.2.

The idea of CbW is to shift the driver interaction from the stabilization level to the guidance level by means of a maneuver-command-based interaction between the driver and the automation. The conventional and continuous interaction between driver and vehicle at the stabilization level is replaced by an event-based communication by means of maneuvers at the guidance level. Thus, the driver of a CbW vehicle delegates the tasks of trajectory planning and vehicle stabilization to the automation, as shown in Fig. 6.3 using the adapted Donges model.

The abstract CbW system architecture shown in Fig. 6.3 builds the basis for a further concretization of the driver-vehicle interaction and the cooperative vehicle guidance concept. Figure 6.4 shows in more detail the elementary CbW system architecture at the guidance level that is divided into the maneuver level and the trajectory level. The driver assigns maneuver commands using the maneuver interface, which also allows the parameterization of the chosen maneuvers—as for example by choosing the excentricity within the lane—and interaction at the stabilization level if desired. The driver's maneuver commands are transferred to the maneuver control that—depending on the actual driving situation—assigns, coordinates, and parameterizes the driver's maneuver command to a pair of exactly one longitudinal

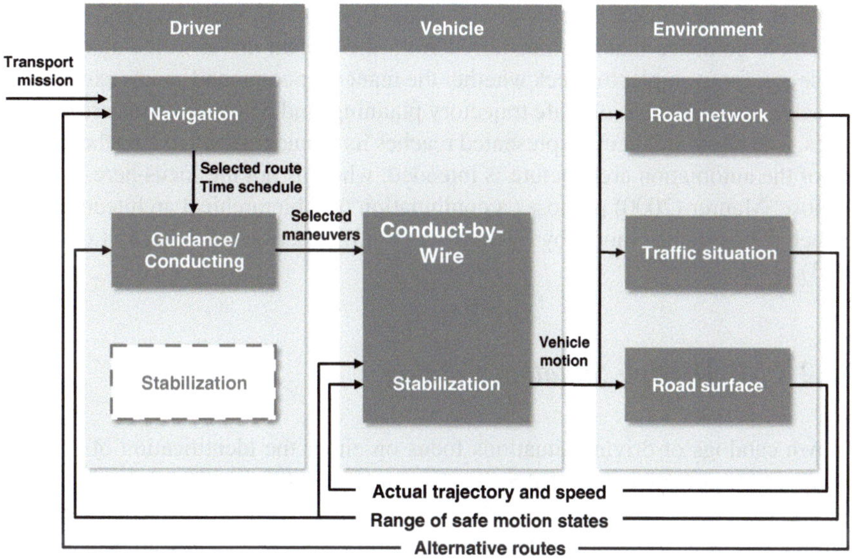

Fig. 6.3 Maneuver-based vehicle guidance (Winner and Heuss 2005, p. 23)

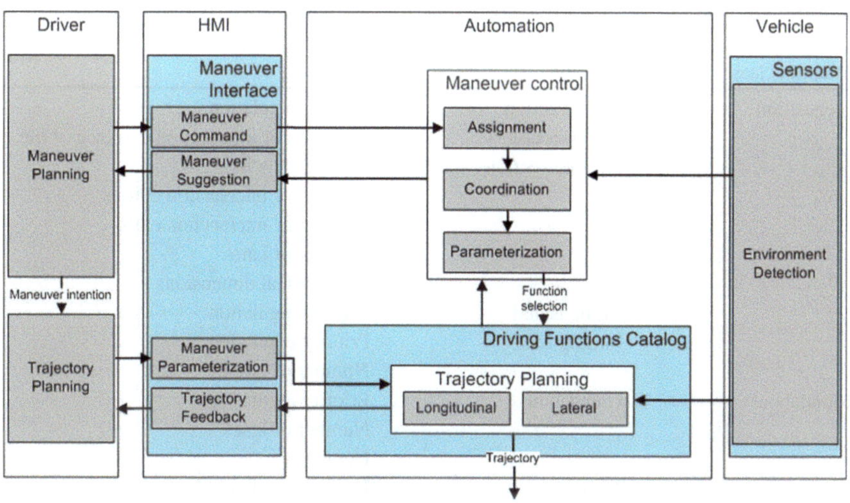

Fig. 6.4 Elementary Conduct-by-Wire system architecture at the guidance level; HMI: Human Machine Interface

and exactly one lateral driving function (Hakuli et al. 2010). These driving functions calculate the trajectories and control the actuators.

As mentioned above, the illustration of the CbW system architecture based on the Donges model helps to describe the different interaction levels between driver and automation as well as the task sharing between these two cooperating partners.

However, the execution of a driver's maneuver command implies several subtasks of vehicle guidance that are transferred from the human driver to the automation. These are, for example, to check whether the maneuver command can be executed in the actual driving situation, safe trajectory planning, and detection of unsafe driving states. The CbW architecture presented reaches its boundaries when a further detailing of the automation architecture is intended, which is not the focus here. For this purpose, Maurer (2000) proposes a combination of a hierarchical architecture with the behavior-oriented model by Rasmussen (1983), described in detail by Lotz (Lotz 2013).

6.2.2 CbW Driving Situation Catalog

Known catalogs of driving situations focus on either the identification of relevant situations for potential new ADAS (e.g. Kuehn et al. 2009) or driver behavior analysis (e.g. Fastenmeier and Gstalter 2007). Both approaches do not present a preferred solution for the analysis of the technical feasibility of CbW, because all potential situations a CbW vehicle has to cope with have to be considered. Therefore, a catalog

Table 6.1 Structure of the Conduct-by-Wire catalog of driving situations (Geyer et al. 2012b, p. 3)

Situation class	Situation	Parameter
Intersection	X-intersection T-intersection Star intersection	Priority regulation Direction of intersection entrance of the ego-vehicle Number of intersection entries Number of intersection exits Number of lanes Geometrical dimensions
	Traffic island	Priority regulation
	Turning road	Priority regulation Number of lanes
Roundabout	Roundabout	Priority regulation Number of lanes Geometry
	Bypass	Priority regulation
Cross traffic	Crosswalk Bicycle lane Railroad crossing	Priority regulation
Parallel traffic	End of lane	One-sided/both-sided Right/left
	Obstacle evasion	Right/left W/wo lane change Parallel/oncoming lane
	Restricted lane	Right/left side
	Traffic lights	

of driving situations (shown in Table 6.1) is developed, derived from German guidelines for road design (Forschungsgesellschaft für Straßen- und Verkehrswesen 2010) and traffic laws (Federal Ministry of Transport 2010) and verified by systematically analyzing real traffic situations in the Rhine-Main region in Germany. The situations of this catalog can be assigned to one of four situation classes. To limit the number of possible parameter variations, the set of relevant parameters that describe a situation can be adapted to the focus of investigation. As the catalog is based on legal guidelines, only admissible parameter combinations are considered, which again limits the number of analyzed situations (Geyer et al. 2012b, p. 2).

6.2.3 Simulation with IPG CarMaker

The vehicle simulation tool IPG CarMaker (2013) is used as first test environment for CbW functions (Hakuli et al. 2010). Besides a completely parameter-based and validated vehicle model, a driver, and a road network model, CarMaker also offers the possibility to integrate the CbW driving functions realized in Matlab/Simulink. Thus, it is possible to design completely reproducible test scenarios in a virtual environment of increasing complexity. Moreover, a systematic parameter variation can be performed using the functionality of the CarMaker TestManager.

6.3 Cooperative Driver-Vehicle Interaction

6.3.1 The Gate Concept

For autonomous vehicle guidance, interaction between driver and automation mainly occurs at navigation level. The driver completely assigns the subsequent steps of vehicle guidance to the automation, which has to guarantee safe accomplishment of the driving mission. Therefore, automation has to make every decision by itself without relying on the driver, who, according to the concept of autonomous driving, is not in the loop. This implies high performance in terms of the perception and cognition abilities of the environment perception system. While interpreting a driving situation, two potential error types have to be avoided: false positive (also known as Type I error or alpha error (Vogt and Johnson 2011, p. 7)) and false negative (also known as Type II error or beta error (Vogt and Johnson 2011, p. 27)). Whereas for false positives automation detects an object where in reality there is none, which might cause unsuitable actions in the current situation, for false negatives automation fails to detect an existing object and thus suitable actions might be missing. The challenge in developing the system architecture is that measures to reduce the probability of occurrence of one type of error automatically increase the probability for the other type. System architectures for autonomous vehicles are mostly based on a complex safety concept as described by Hörwick and Siedersberger (2010) and on conservative decision strategies admitting more false positives. Hereby, autonomous error handling strategies become quite time-consuming because different alternatives with increasing potential risk have to be assessed sequentially (Urmson et al. 2008, p. 451).

This decision process could be significantly shortened by integrating the human driver whose abilities in deciding whether a positive detection is true or false are superior to those of the automation (Winner and Weitzel 2012, p. 664). Moreover, the human driver is able to analyze complex driving situations and if necessary to derive alternatives. Thus the integration of the human driver is the motivation for cooperative vehicle guidance concepts in order to increase the controllability of these highly automated concepts. Interaction between driver and automated systems and the analysis of the system behavior are often considered from an ergonomic point of view, concentrating on the driver, analyzing his/her abilities to understand and predict the behavior of the automated system and his/her capabilities of intervening correctly in case of reaching system borders. Even though this is a key factor for a successful cooperation between driver and automation, the relevance of a clearly defined system behavior for the vehicle-related design of the overall system architecture is mostly neglected. Moreover, the technical requirements of the automation concerning the driver-vehicle interaction have to be considered (Geyer et al. 2011, p. 2).

As described above, the CbW principle is based on a clear task assignment between the driver (maneuver command) and the automation (maneuver execution and optional maneuver suggestion). Depending on the driving functions executed and the current driving situation, the elementary CbW system architecture shown in Fig. 6.4 reaches its boundaries when facing situations with increased collision risk due to crossing other priority trajectories. These driving situations require that additional information be taken into account within the decision process in order to guarantee a safe continuation of the driving mission. In Geyer et al. (2011) a new approach for the systematic analysis of information required for decision-making in different driving scenarios has been introduced. The key element of this approach is the "gate concept" that consists in a segmentation of the vehicle guidance task. The gates mark the points along a planned trajectory where a decision has to be made on continuing the execution of a driver's maneuver command and thus of the driving mission. Each gate is assigned to an information cluster, which comprises different information needed at that point.

In Fig. 6.5, the gate concept is demonstrated for an exemplary driving scenario. The CbW vehicle enters an X-intersection where the priority regulation is "yield-to-right" from the direction South and turns left. During the execution of this maneuver, a sequence of two gates has to be passed. The first gate "Intersection entry (I,E)" is positioned before entering the intersection area. For the decision on whether a safe continuation of the maneuver execution to the next gate is possible the vehicles with higher priority that enter the intersection from direction East have to be considered. Moreover, it has to be checked whether the area between the two gates is occupied. For passing the second gate "Intersection left (I,L)", oncoming vehicles from direction North and the area to the intersection exit are taken into account for decision-making.

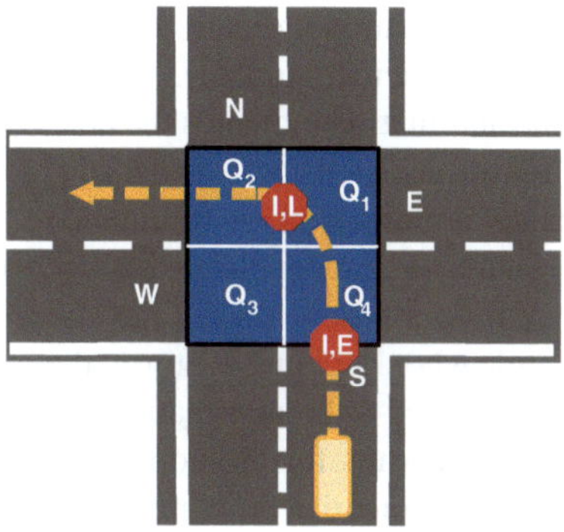

Fig. 6.5 Exemplary driving scenario with identified gates

In cases where no decision is made by the driver or the automation, the gate stays, metaphorically speaking, closed. In that case, the maneuver currently being executed is put on hold and replaced by a gate-approaching maneuver. Common to all approaching strategies is that they stop the vehicle at the position of the gate according to the CbW safety concept. Once the gate is "unlocked", the stored maneuver is reactivated from the memory. The gate concept needs the maneuver management to be added to the elementary CbW system architecture at the guidance level, as shown in Fig. 6.6.

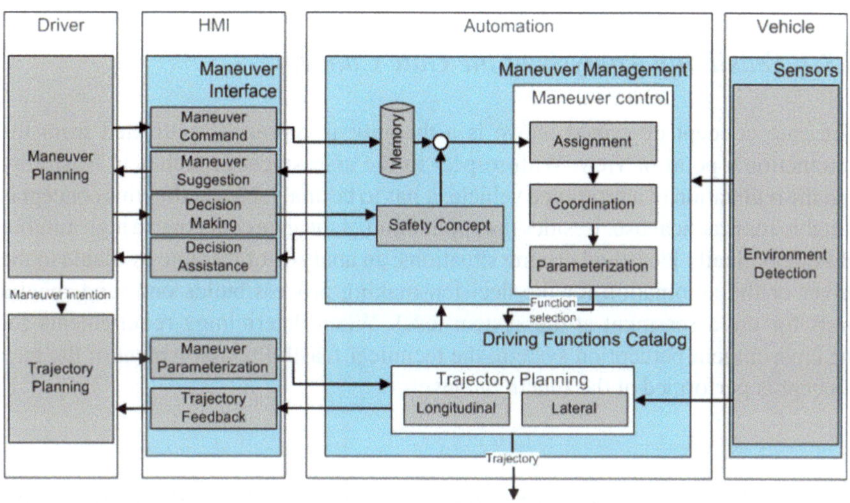

Fig. 6.6 Extended Conduct-by-Wire system architecture at the guidance level (Geyer et al. 2012a, p. 4)

When approaching a gate, the driver of a CbW vehicle can be cooperatively assisted by the automation in decision-making at different automation levels (Geyer et al. 2011, p. 3):

- **Automation level 1:** The automation indicates the next upcoming gate; the decision is made by the driver (explicit command).
- **Automation level 2:** The automation indicates the next upcoming gate and makes a suggestion and waits for the decision command made by the driver (confirmation of suggestion or decline by another explicit command).
- **Automation level 3:** Share of responsibility between driver and automation, depending on the potential error when making the decision. For a potential false positive error—for example when deciding to stop at the intersection entry because the intersection area is occupied—the automation indicates the next upcoming gate, makes a suggestion and waits for the confirmation (or alternative explicit command). Decisions with a potential false negative error—for example when deciding to pass the intersection because the intersection area is not occupied—the automation indicates the next upcoming gate and executes the maneuver while informing the driver about the decision and future action.

The described decision assistance leads to an extension of the functionality of the maneuver interface, shown in Fig. 6.6. It is important to note that these automation levels represent different system characteristics of CbW. As shown in the following sections, the technical requirements increase with increasing automation level. Hence, a stepwise realization of the CbW concept in a prototype vehicle should follow these automation levels. Thus, the functionality of the automation changes for each stage of development.

6.3.2 Feasibility Analysis of the Gate Concept

The gate concept described above is a theoretical approach, motivated from the automation's point of view. With respect to the assessment of technical feasibility and the realization of a prototype vehicle, it has to be analyzed whether this concept is suitable for practical use. Besides the application of the gate concept to a high number of systematically identified driving situations, an analysis of the time available to the driver or the automation for the decision-making process builds one fundamental basis for the assessment at the system level. When determining requirements for the environment perception system, the technical feasibility assessment of the gate concept is performed at the functional level.

6.4 Feasibility Analysis of the Gate Concept at the System Level: Temporal Analysis

The temporal analysis of the gate concept follows a four-step approach described below (Geyer et al. 2012b).

6.4.1 Identification of Gate Sequences

The goal of this first step is to apply the gate concept to a high number of driving situations, which are systematically derived from the catalog shown in Table 6.1. An analysis of the gate sequence is performed for each combination of driving situations and possible driving maneuvers.

6.4.2 Parameter Analysis

The next step consists in identifying the parameters that influence the time available for decision-making. These are the

- position of the gates
- velocity
- approaching strategy
- time of gate unlock.

The positioning of the gates depends on the geometrical representation of the considered driving situation and the planned trajectory. Parameters that describe the driving situation might, for example, be the relative position of the connected roads of an intersection, the width of the traffic lanes or the positioning of crosswalks. These parameters are chosen in accordance with real traffic situations.

Another parameter that has an impact on the time available for decision-making is the velocity at which the vehicle approaches the gate and the approaching strategy. While the former primarily depends on the velocity chosen by the driver v_{set}, the latter depends on the acceleration performance of the vehicle a_V and the deceleration assigned to the gate.

Finally, the time when the driver or the automation decides to "unlock" a gate influences the velocity with which the vehicle passes the gate, and thus the time that is available to make a decision on the continuation of the driving mission at the following gate. Three different cases can be distinguished, exemplarily shown for two consecutive gates in Fig. 6.7 assuming constant acceleration. The left side shows the velocity-displacement courses. By plotting the square of the velocity over displacement, constant accelerations are shown as straight lines with the acceleration as slope. The right side shows the same courses over time.

In case I, the first gate is unlocked before the approaching strategy of the first gate is executed. The velocity at the entrance of the first gate v_{GE} depends on the chosen velocity v_{set}. Moreover, v_{GE} is automatically limited to the maximum velocity $v_{GE,max}$ as a function of the distance between the two gates Δs_G and the deceleration D_{G2} assigned to the approaching strategy of gate 2. This means that the deceleration for gate 2 begins before (I.a) or after (I.c) gate 1 is passed or at gate 1 (I.b) respectively. In case II the first gate is unlocked after the vehicle has stopped. The velocity between the two gates v_{BG} up to which the vehicle accelerates after gate 1 is unlocked depends

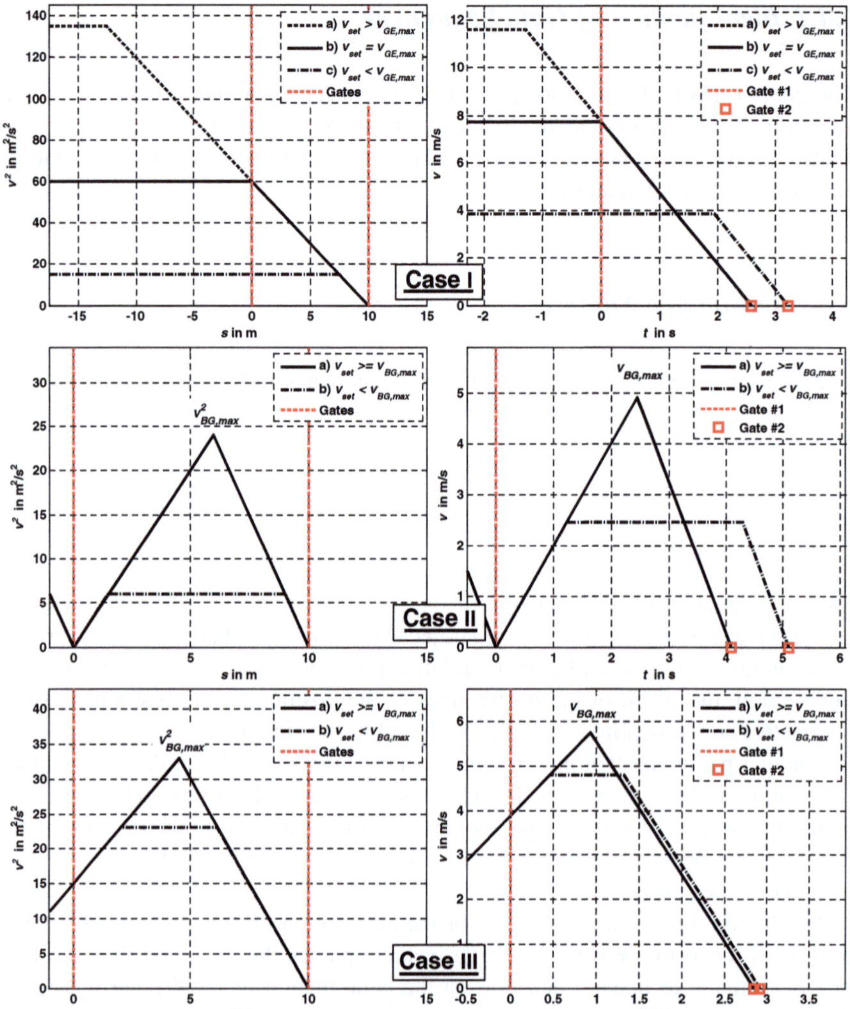

Fig. 6.7 Possible cases for the time of gate unlock; v: velocity, t: time, s: distance

on v_{set} and is limited to $v_{BG,max}$, that depends on Δs_G, D_{G2}, and the acceleration performance of the vehicle a_V.

In case III, v_{GE} is lower than $v_{set,}$ which can be the case when the decision to pass gate 1 is made after the deceleration for gate 1 has started or when the distance to the gate is too short to reach either v_{set} or $v_{GE,max}$. Thus, the cases I and II represent the extremes of case III. The time available between the two gates primarily depends on the velocity at the first gate v_{GE}. The cases where the acceleration performance and the distance to gate 1 are sufficient to reach either v_{set} or $v_{GE,max}$ are similar to case I. If $0 < v_{GE} \leq v_{GE,max}$, v_{BG} depends on v_{set} and is limited to $v_{BG,max}$ as a function of Δs_G, D_{G2}, a_V, and v_{GE}. The influence on the time available Δt_G between the two gates is qualitatively shown on the right side of Fig. 6.7.

6.4.3 Simulation

To describe the third step, exemplary simulation results are given for a scenario where the CbW vehicle turns left at an X-intersection with traffic lights while a tram lane and two crosswalks have to be passed, as shown in Fig. 6.8. Because the area in front of the second crosswalk does not allow a safe stop of the vehicle, the decision to pass this crosswalk is made at gate 3. Thus, the resulting gate sequence consists of three gates with a relative distance of 20.9 and 9.2 m.[1]

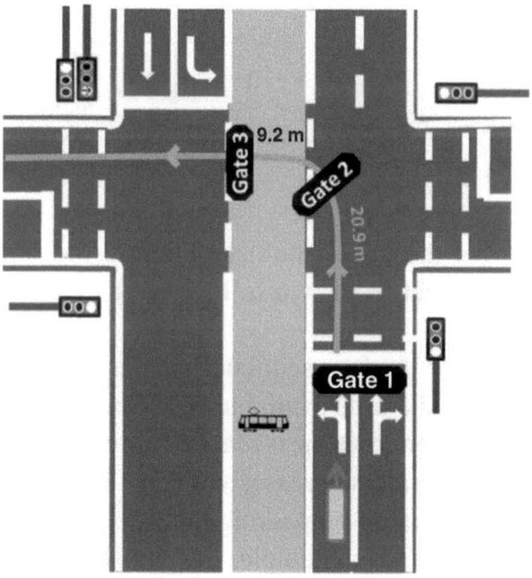

Fig. 6.8 Example for passing the three gates assigned to the turning left maneuver at an X-intersection (Geyer et al. 2012b, p. 4)

As mentioned above, there are different time-influencing parameters. To demonstrate this method, only two parameters have been modified, namely the deceleration D_G(3 and 5 m/s^2) that is the same for all gates of a sequence and the time when the gates are unlocked. The constant simulation parameters are shown in Table 6.2.

Based on the three different cases when a gate can be unlocked shown in Fig. 6.7, there are $3^3 = 27$ variations for a sequence of three gates. This number can be reduced to four sequences (shown in Table 6.3) when taking into account that one case for two consecutive gates is represented in several sequences. The indication "stopped" means that the gate is unlocked after the vehicle has stopped at the gate,

[1] The geometrical dimensions are taken from an intersection in Darmstadt, Germany [49°52'03.42 "N 8°39'50.28" E]

Table 6.2 Constant simulation parameters

Parameter	Variable	Value
Distance between the gates	Δs_{G12}	20.9 m
	Δs_{G23}	9.2 m
Velocity	v_{set}	50 km/h
Maximum acceleration of the vehicle	a_v	$2\,\text{m/s}^2$

Table 6.3 Variation of the time when gates are unlocked

Sequence	Gate 1	Gate 2	Gate 3
1	stopped	stopped	stopped
	Case II		Case II
2	stopped	passed	stopped
	–		Case III
3	passed	passed	stopped
	–		Case I
4	passed	stopped	stopped
	Case I		Case II

while “passed” means that the gate is unlocked before the approaching strategy is initiated. Moreover, the sequence of two following gates is assigned to one of the three cases of Fig. 6.7.

The simulation results for the four sequences and a variation of the deceleration D_G are shown in Table 6.4. The time available between two gates varies from 1.8 to 5.7 s, while, as expected, the longest time is reached for case II where the vehicle stops at each gate and the time generally decreases with D_G. The results put in parentheses are those cases where the decision on passing the gate is made before reaching that gate. In the case of sequence 3 for example, the decision has been possible before reaching gate 2 and thus the time between gate 1 and gate 2 is not critical. Although most of these early decision cases can be neglected for the criticality analysis, some of them are relevant as they influence the subsequent two gates. Moreover, these results might be of interest when analyzing the time available for changing a made decision. Depending on the actual situation, this time may possibly be useful for making a decision on the subsequent gate.

Table 6.4 Simulation results (Geyer et al. 2012b, p. 5); D: deceleration, t: time

	$D_G = 3\,\text{m/s}^2$		$D_G = 5\,\text{m/s}^2$	
	Δt_{G12} in s	Δt_{G23} in s	Δt_{G12} in s	Δt_{G23} in s
Sequence 1	5.7	3.8	5.3	3.6
Sequence 2	(4.6)	2.3	(4.6)	**1.8**
Sequence 3	(1.9)	2.4	(1.6)	**1.8**
Sequence 4	3.6	3.7	2.8	3.5

6.4.4 Result Evaluation

As a final step, the calculated time available between two consecutive gates has to be compared with the time need of a human driver or the automation.

Many different studies can be found in literature concerning the time between perception and reaction of a human driver, which is the focus here. This time varies between 0.5 and 3.5 s or is sometimes even above this value (Olson and Faber 2003, p. 315). Of course, these studies are only valid for conventional vehicle guidance and cannot simply be transferred to the CbW concept. However, they allow a rough estimation of a driver's time need for decision-making, which is set at the mean value of 2.0 s for the result evaluation made here.

Moreover, only those gate sequences are analyzed where a need for decision-making exists. For the exemplary scenario, critical times—marked in bold in Table 6.4—occur in sequence 2 and sequence 3 between the gates 2 and 3 for $D_G = 5\,\mathrm{m/s^2}$. For all other parameter variations, the simulated times are not critical.

6.4.5 Conclusion

The time analysis presented above is a fundamental contribution to the assessment of the gate concept and thus of the technical feasibility of Conduct-by-Wire. The goal of this analysis at the system level is to follow a worst-case approach. Thus, a simple control strategy with constant accelerations and linear velocity curves has been chosen. The results of this analysis–that reveals situations that are identified as being time critical—can be used as input for the next development step at the functional level. There are two promising approaches to follow. The first is the transfer of the human driver behavior at decision points to the behavior of the CbW automation. The second is the combination of the gate concept with an adaptive approaching strategy (Geyer et al. 2012b, p. 5) that generates the time needed for decision-making.

6.5 Feasibility Analysis of the Gate Concept at the Functional Level: Determination of the Requirements for the Environment Perception System

In this section, the feasibility assessment of the gate concept at the functional level is demonstrated for the analysis of the environment perception system, which provides the information needed at the gates in order to implement decision assistance for different automation levels as described in Sect. 6.3.1. The technical requirements are systematically derived following a top-down approach based on the analysis of the decision process and compared with present-day sensor specifications. The focus lies on machine perception and maps as information sources, while Car2X-solutions are not considered.

6.5.1 Analysis of the Decision Process

The decision process during the gate approach can be divided into different steps, as shown in Fig. 6.9. The starting point is the state of decision-free driving, where the vehicle guidance is realized by an activated longitudinal and lateral driving function.

The decision-free driving ends in the event of a changing situation that requires additional decisions concerning the continuation of the driving mission—for example when approaching a roundabout or a static obstacle within the own lane. In order to detect this state transition, the automation needs to be able to identify those criteria that allow a falsification of the basic hypothesis of decision-free driving. Moreover, for the highest CbW automation level, the situation knowledge is needed for checking whether the driver's maneuver command can safely be executed in the current situation. As this automation level is like fully-automated driving in regard to the maneuver execution, one cannot assume that the driver is fully attentive in every situation—in contrast to the other automation levels that require an action by the driver. Thus, it might be possible that the driver is not aware of decision errors made by the automation, and this presents a safety risk.

The first step in gate localization is made by the gate identification, consisting of a logical assignment of the parameters that describe the current driving situation, the definitions of the gate positions as well as the information needs at these gates. Thus, this step does not present any additional requirements concerning the environment perception. In the second step, the gate positions are determined based on the current

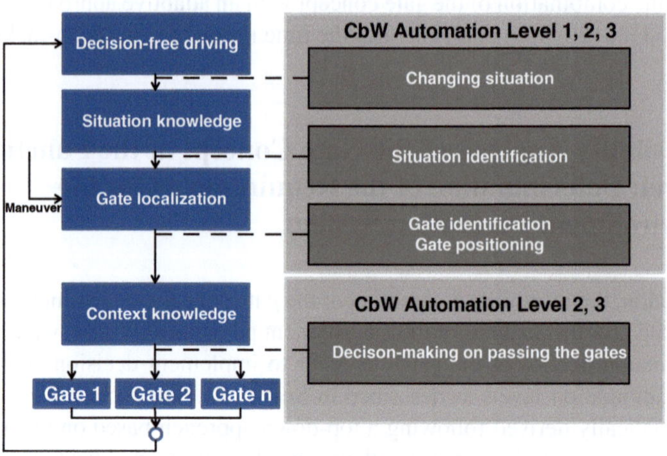

Fig. 6.9 Decision process during the gate approach (Geyer et al. 2012b, p. 8); CbW: Conduct-by-Wire

surroundings data, e.g. detected road markings. Thus, the "situation knowledge" and the "gate identification" allow the localization of the gates that have to be passed for a driver's maneuver command. This builds the basis for all three CbW automation levels.

The CbW automation levels 2 and 3—where the automation makes a suggestion or a decision about the gate passage—require additional "context knowledge" about the current scenario in order to be able to cover the information needs for decision-making during the gate approach or at the gate. One example for this context knowledge might be to have information on the trajectory of objects following a traffic circle.

Each of the three steps of the decision process—situation knowledge, gate localization, and context knowledge—leads to specific requirements concerning the environment perception. Section 6.5.2 briefly describes how these requirements can be determined for an exemplary intersection scenario, a more detailed description can be found in (Geyer et al. 2012a).

6.5.2 Requirement Determination

6.5.2.1 Situation Knowledge

The information need to detect a state transition that requires a decision is covered by identifying the parameters that describe a situation. For the given example of detecting an intersection, this might be the detection of road markings, constructional boundaries, traffic signs or traffic lights.

As described above, the automation transfers the vehicle to a safe state by stopping at the gate if no decision is made by the driver or the automation. The distance to the gate at which the approaching maneuver is initiated depends on the velocity and the acceleration assigned to the gate, approximately determined by

$$s_{AS} = -\frac{v_{Ego}^2}{2D_G} \tag{6.1}$$

For an exemplary deceleration with 3 $\mathrm{m/s^2}$ and an initial velocity of 50 km/h, the distance to the gate is 63 m. This means that the end of decision-free driving has to be detected at the latest at that distance to the first gate of a situation, which then corresponds to the minimum longitudinal sensor range.

6.5.2.2 Gate Localization

Figure 6.10 shows the possible gates at an intersection for the maneuvers "turn left", "turn right", "driving straight through the intersection", and "u-turn". The position of the gate I, R depends on the planned trajectory of the turning right maneuver.

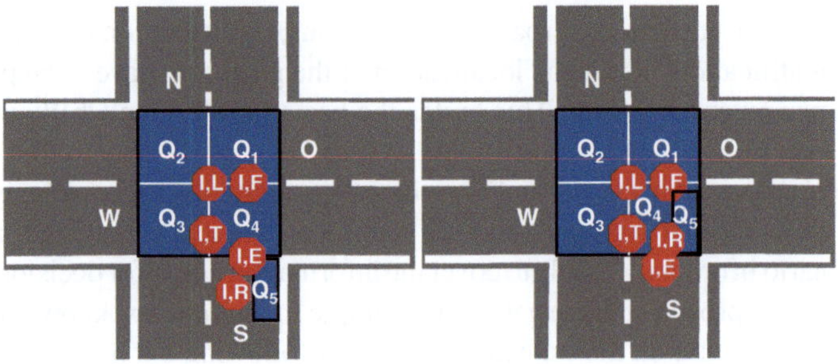

Fig. 6.10 Intersection with occupancy map ($Q_1 - Q_5$), entry directions of other traffic members, and gates (I, x) (Geyer et al. 2012a, p. 10)

For the gate positioning, real (for example lane markings) and virtual (for example the connection of interrupted lane markings) boundary lines have to be detected or identified. This information need has to be covered at the same time as the situation identification. When a gate sequence has to be passed during a scenario, the information need for passing a gate is required at the position of the previous gate at the latest.

6.5.2.3 Context Knowledge

The context knowledge allows the automation to make a proposal or a decision on the gate passage. For the intersection example, the required information is the signal of potential traffic lights, other approaching traffic members with priority and the occupancy of intersection quadrants that have to be passed during the maneuver execution. The determination of the required sensor range is made for two extreme cases. The first case is the situation where the decision on passing the gate is made after the vehicle has stopped at the gate. The second case is the situation where the maneuver execution is not interrupted by the gate-approaching maneuver.

The required minimum sensor range is determined based on the required time for the maneuver execution t_{ME}. A safety corridor is defined, that other traffic members are not allowed to enter during the maneuver execution of the CbW vehicle. Figure 6.11 shows the safety corridor for an intersection scenario. For safe maneuver execution, a safety criterion is defined that is based on the time-to-safety-corridor (*TTS*). This time denotes the time needed for reaching the safety corridor at a distance d_S with constant velocity v_{Object}. The *TTS* of approaching traffic members has to correspond at least to the required time for the maneuver execution t_{ME}:

$$TTS = \frac{d_S}{v_{Object}} \geq t_{ME} \tag{6.2}$$

The minimum sensor ranges for the coordinate system of the ego vehicle (shown in Fig. 6.11) is calculated based on the determined maneuver execution time t_{ME}, the

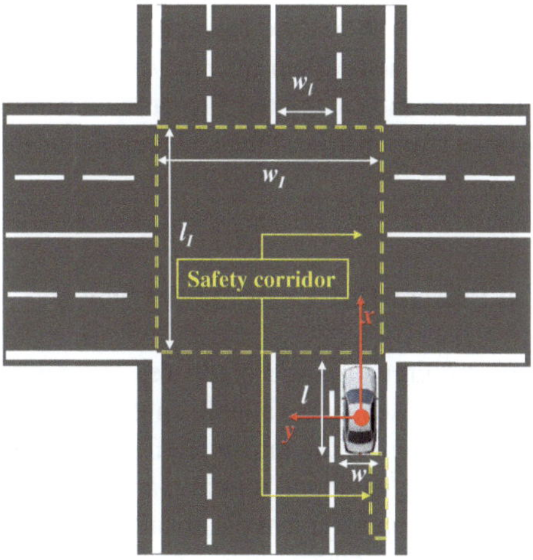

Fig. 6.11 Safety corridor at an intersection (Geyer et al. 2012a, p. 14)

distances to the end of the safety corridor $d_{s,x}$ and $d_{s,y}$, and the velocities of both the ego vehicle v_{Ego} and approaching traffic members v_{Object}:

$$R_x = v_{Object} \cdot t_{ME} + v_{Ego} \cdot t_D + d_{S,x} \tag{6.3}$$

$$R_y = v_{Object} \cdot t_{ME} + v_{Ego} \cdot t_D + d_{S,y} \tag{6.4}$$

The time t_{ME} comprises the time that is required for the maneuver execution—e.g. turning left at an intersection—as well as the time the machine perception needs for the object detection. The latter is estimated at $t_D = 1\text{s}$.

Using simple trigonometric functions, these ranges can be transferred to typical sensor specifications of range-based sensors, i.e. the maximum sensor range and the

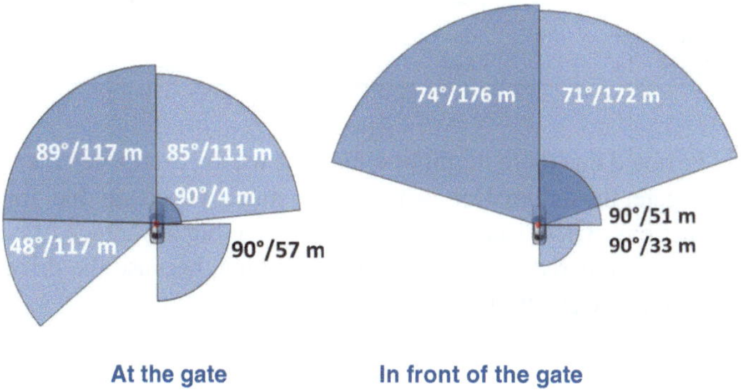

Fig. 6.12 Determined minimum sensor ranges

maximum azimuth. Figure 6.12 shows the calculated sensor requirements based on different assumptions for a typical urban intersection scenario.[2] For camera-based systems it is the resolution of the calculated *TTS* at the determined distances R_x/R_y (which depend on the object dimensions, the focal length, and the measurement time) that is relevant.

6.5.2.4 Assessment of Present-Day Environment Sensors

The determined sensor requirements are compared to sensor principles common today and the specifications of sensors available on the market. A qualitative assessment is shown in Table 6.5. To cover the information needs derived from the situation knowledge, the camera and the extended digital map fulfill the requirements. However, both information sources have typical disadvantages. For digital maps these are the currentness of the map data and the precision of positioning when using GPS. For the camera, the disadvantage is the limited range, which is not sufficient at higher velocities. A promising solution to this problem might be the combination of a simple digital map—that, for example, allows a rough estimation of the distance to the next gate—with camera data that cover the residual information need and that can be used to verify the map data. A similar estimation is made for the gate localization, although the digital map can contribute only little information. To cover the context knowledge, radar and lidar present the preferred solutions, while lidar sensors have the advantage of a higher azimuth and a better performance in detecting pedestrians and cyclists. The most challenging requirement is to cover the area on the right side of the CbW vehicle.

Table 6.5 Assessment of different information sources (Geyer et al. 2012a, p. 22)

Information needs	Environment sensors					Maps	
	Radar	Lidar	Camera	3D-ToF	Ultrasonic	Digital map	Extended digital map
Situation knowledge	–	–	+	–	–	(+)	+
Gate localization	–	+	+	+	–	(+)	(+)
Context knowledge	+	+	(+)	(+)	–	–	–

It should be noted that there might be situations where, even when the vehicle has stopped at the gate, the automation is not able to make a decision due, for example, to sight obstructions. These cases especially show the big advantage of cooperative vehicle guidance compared to fully-automated driving, because the driver is in the loop and can intervene at any time.

[2] Assumptions: intersection dimensions= 14×14 m; $v_{Ego} = 50$ km/h; $v_{Object} = 70$ km/h; $a_V = 2\,m/s^2$; $D_{Ego} = -3\,m/s^2$; $a_{y,max} = 4\,m/s^2$

6.6 Conclusion and Outlook

This chapter demonstrates the technical feasibility assessment of an innovative vehicle guidance concept based on the Conduct-by-Wire principle. The assessment follows a systematic top-down approach that allows an assessment in the early concept phase. The basis of all subsequent development steps is built by the system architecture. The latter defines the system functionality, the system limits, and the interaction of the different system elements. This allows for early investigations to be made, such as the time analysis of the gate concept. These studies at the system level form the basis for concretizing the functional development, as shown for the determination of the requirements of the environment perception system.

Thus, this systems engineering approach makes it possible to handle the complexity and implement a stepwise development. Furthermore, a decision on whether the realization of this innovative concept is worthwhile can be made at different abstraction levels in an early phase before a prototype system has to be built. The results presented show some of the future technical challenges, while initial promising investigations with test persons have demonstrated the acceptance of maneuver-based vehicle guidance (Kauer et al. 2012).

Acknowledgments This work was supported by the German Research Foundation (Deutsche Forschungsgemeinschaft—DFG) as part of the research project, "Von Conduct-by-Wire zu einer kooperativen Fahrzeugführung: Erweiterung und Validierung eines Konzepts für kooperative, manöverbasierte Fahrerassistenz" (WI 3153/2-2).

References

Ardelt, M., Coester, C., Kaempchen, N.: Highly automated driving on freeways in real traffic using a probabilistic framework. IEEE Trans. Intell. Transp. Syst. **13**(2), 1–10 (2012). Institute of Electrical and Electronics Engineers, New York

Donges, E.: Aspekte der aktiven Sicherheit bei der Führung von Personenkraftwagen, Automobil-Industrie, vol. 2, pp. 183–190. Vogel Business Media GmbH & Co. KG, Würzburg (1982)

Fastenmeier, W., Gstalter, H.: Driving task analysis as a tool in traffic safety research and practice. Saf. Sci. **45**(9), 952–979 (2007)

Federal Ministry of Transport, Building and Urban Development: Straßenverkehrs-Ordnung, Berlin (2010)

Flemisch, F.O., Adams, C.A., Conway, S.R., Goodrich, K.H., Palmer, M.T., Schutte, P.C.: The H-Metaphor as a Guideline for Vehicle Automation and Interaction, NASA/TM-2003-212672. Langley Research Center, Hampton (2003)

Forschungsgesellschaft für Straßen- und Verkehrswesen FGSV: Richtlinien für die Anlage von Stadtstraßen RASt 06. FGSV, Cologne (2010)

Geyer, S., Hakuli, S., Winner, H., Franz, B., Kauer, M.: Development of a cooperative system behavior for a highly automated vehicle guidance concept based on the Conduct-by-Wire principle. In: IEEE Intelligent Vehicles Symposium, 2011, 5–9 June Baden-Baden, Germany (2011)

Geyer, S., Hakuli, S., Winner, H., Franz, B., Kauer, M.: Ermittlung der Anforderungen an die Umfelderkennung für Conduct-by-Wire, 5. Tagung Fahrerassistenz. Munich, Germany, 15–16 May 2012a

Geyer, S., Hakuli, S., Winner, H., Franz, B., Kauer, M.: Temporal Analysis of the Gate Concept as Enabler for Highly Automated Driving based on the Conduct-by-Wire Approach. In: IEEE Intelligent Transportation Systems Conference, 2012. Anchorage, AK, USA, 16–19 Sept 2012b

Hakuli, S., Kluin, M., Geyer, S., Winner, H.: Development and validation of manoeuvre-based driver assistance functions for Conduct-by-Wire with IPG CarMaker. In: FISITA 2010 World Automotive Congress, Budapest, Hungary 30 Mai–4 June 2010

Hörwick, M., Siedersberger, K.: Strategy and Architecture of a Safety Concept for Fully Automatic and Autonomous Driving Assistance Systems. In: IEEE Intelligent Vehicles Symposium, 2010. San Diego, CA, 21–24 June 2010

Hummel, T., Kühn, M., Bende, J., Lang, A.: Advanced Driver Assistance Systems. An investigation of their potential safety benefits based on an analysis of insurance claims in Germany. German Insurance Association Insurers Accident Research, Research report FS 03, Berlin (2011)

IPG CarMaker: http://www.ipg.de

Kauer, M., Franz, B., Schreiber, M., Bruder, R., Geyer, S.: User acceptance of cooperative maneuver-based driving—a summary of three studies, in Work: A Journal of Prevention, Assessment and Rehabilitation 41(1), 4258–4264 (2012), IOS Press, Amsterdam

Kuehn, M., Hummel, T., Bende, J.: Benefit estimation of advanced driver assistance systems for cars derived from real-life accidents. In: 21st International Technical Conference on the Enhanced Safety of Vehicles ESV 2009. Germany, Stuttgart, 15–18 June 2009

Lotz, F.: System architectures for automated vehicle guidance concepts. In: Maurer, M., Winner, H. (eds.) Automotive Systems Engineering, pp. 1–23. Springer, Heidelberg (2013)

Maurer, M.: Flexible Automatisierung von Straßenfahrzeugen mit Rechnersehen, Reihe 12, Verkehrstechnik, Fahrzeugtechnik, vol. 443. VDI, Düsseldorf (2000)

Olson, P.L., Faber, E.: Forensic Aspects of Driver Perception and Response. Lawyers & Judges Publishing Company, Tucson (2003)

Rasmussen, J.: Skills, rules, and knowledge, signals, signs, and symbols, and other distinctions in human performance models. IEEE Trans. Syst. Man Cybern. **SMC-13**(3), 257–266 (1983), Institute of Electrical and Electronics Engineers, New York

Saust, F., Wille, J.M., Lichte, B., Maurer, M.: Autonomous vehicle guidance on Braunschweig's inner ring road within the Stadtpilot project. In: IEEE Intelligent Vehicles Symposium, 05–09 June 2011. Baden-Baden, Germany (2011)

United Nations: Convention on Road Traffic and Road Signs. Austria, Vienna, 8 Nov 1968

Urmson, C., Anhalt, J., Bagnell, D., Baker, C., Bittner, R., Clark, M.N., Dolan, J., Duggins, D., Galatali, T., Geyer, C., Gittleman, M., Harbaugh, S., Hebert, M., Howard, T.M., Kolski, S., Kelly, A., Likhachev, M., McNaughton, M., Miller, N., Peterson, K., Pilnick, B., Rajkumar, R., Rybski, P., Salesky, B., Seo, Y., Singh, S., Snider, J., Stentz, A., Whittaker, W., Wolkowicki, Z., Ziglar, J., Bae, H., Brwon, T., Demitrish, D., Litkouhi, B., Nickolaou, J., Sadekar, V., Zhang, W., Struble, J., Taylor, M., Darms, M., Ferguson, D.: Autonomous driving in urban environments: Boss and the urban challenge. J. Field Rob. **25**(8), 425–466 (2008)

Vogt, W.P., Johnson, R.B.: Dictionary of Statistics and Methodology. SAGE Publications, Inc., Los Angeles (2011)

Winner, H., Heuss, O.: X-by-Wire Betätigungselemente–Überblick und Ausblick. In: Winner, H., Landau, K. (eds.) Darmstädter Kolloquium Mensch und Fahrzeug 2005, Cockpits für Straßenfahrzeuge der Zukunft. Ergonomia, Stuttgart (2005)

Winner, H., Weitzel, A.: Quo vadis, FAS? In: Winner, H., Hakuli, S., Wolf, G. (eds.) Handbuch Fahrerassistenzsysteme, pp. 658–667. Vieweg+Teubner Verlag, Wiesbaden (2012)

Part III
Functional Safety

Chapter 7
Objective Controllability Assessment for Unintended ADAS Reactions

Alexander Weitzel

7.1 Introduction

Advanced Driver Assistance Systems (ADAS) are developing rapidly. They rely on the environment perception system to assist and support the driver to avoid accidents. This development urges a need for test methods and test tools for the assessment of different aspects of these systems. Beside verification of the functions, it is also necessary to validate the positive effects of these systems on traffic safety. To identify the requirements for these more global evaluations, the whole system of the vehicle, the driver and the environment, has to be taken into account.

Existing testing methods and testing tools for ADAS have a varying spectrum of purposes. Many of these methods are designed to verify the sensor detection abilities or the performance of the hazard identification and warning algorithm of an ADAS function. Their purpose is to prove with measurements that a specific attribute or feature of the product fulfills the defined requirements. Therefore, these testing methods and tools need to be highly specific for the given attribute.

Regarding the purpose of an ADAS system in real traffic situations, the intended benefit is often global in terms that it is inclusive, because the systems boundaries used for the assessment include the driver in the vehicle within the traffic situation. Example intended benefits are "reduce the accident rate" or "reduce the severity of accidents". Moreover, it needs to be proven that the benefits outweigh the potential downsides, like hazards due to failures or false reactions. In case of potential failures, these tests are required to obtain the functional safety approval of the system according to ISO 26262 (2009, Part 3, p. 6) . False reactions are not addressed in this standard, although the resulting situation for the driver may be similar.

These "global" characteristics of a system cannot necessarily be derived from the functional requirements, as they are highly dependent on situations and usage of the

A. Weitzel (✉)
Fachgebiet Fahrzeugtechnik, TU Darmstadt, Petersenstrasse 30, 64287 Darmstadt, Germany
e-mail: weitzel@fzd.tu-darmstadt.de

M. Maurer and H. Winner (eds.), *Automotive Systems Engineering*,
DOI: 10.1007/978-3-642-36455-6_7, © Springer-Verlag Berlin Heidelberg 2013

vehicle. Therefore, test methods for the characteristics need to be highly relevant for the field, must take into account the whole system of the driver with the vehicle in the environment and must allow a transferable assessment which is independent of the specific system.

7.2 Requirements for the Assessment of Global Characteristics

To assess the global characteristics, suitable test situations have to be identified. These situations must clearly reveal the characteristic and must be evaluated and ranked as to their relevance in real traffic. Therefore, the characteristic and its influencing factors need to be analyzed and connected to the situational factors. By ranking the influencing factors and identifying their relevance and impact, the minimum number of test situations will be determinable so that the test effort can be kept to a minimum. The relevance of a situation is highly dependent on the usage of the vehicle. For this reason, the definition of the situation spectrum depicting the use profile is crucial.

Based on the situations, objective values must be defined, describing the attribute in a common and transferable way. For the benefit of collision mitigation systems, for example, Hoffmann (2008, p. 34) uses the reduction of speed before the crash, standing proxy for the reduction of crash energy.

In addition, the identification of an absolute reference for these objective values increases transferability and comparability to similar systems and enables preceding risk assessment methods to be carried out, such as the "Hazard and Risk Assessment" according to ISO 26262 (2009, Part 3, p. 6).

7.3 Controllability Assessment in ISO 26262

An example for a global characteristic of an ADAS system is the controllability of its functions in case of unintended reactions. If the system is working as intended, it will reduce accidents, but relying on an environmental sensor system can also lead to false reactions by the function.

In such cases, the driver or other involved persons must be able to reach a safe state of the vehicle and avoid an accident. This is referred to as "controllability" of the event. For functional safety, ISO 26262 defines requirements for safety processes for electric and electronic systems in the automotive industry. Controllability is therein defined in classes describing the percentages of persons who are able to control a hazardous situation (see Table 7.1).

In combination with the classification of the Exposure and Severity of a potential hazard, ISO 26262 determines an Automotive Safety Integrity Level (ASIL) which defines the minimum safety requirements for hard- and software components of the function. ASIL D, for example, describes the maximum allowable hardware failure to be less than 10^{-8} per operating hour.

Table 7.1 Classes of controllability (ISO 26262 2009, Part 3, p. 9)

Class	C0	C1	C2	C3
Description	Controllable in general	Simply controllable	Normally controllable	Difficult to control or uncontrollable
Definition	Controllable in general	≥99 %	≥90 %	<90 %
		or more of all drivers or other participants are usually able to avoid a specific arm		

The recommended methods and processes for the controllability assessment within the meaning of ISO 26262 are summarized in the "Code of Practice" (PReVENT 2009, p. 13ff). The Code gives as an example that for proving controllability at level C2 by a car clinic with naïve test subjects, a minimum of 20 valid data sets with positive controllability is needed, resulting in a minimum of 85 % of controllability, which is considered to be sufficient for C2. With the same approach and at the same level of confidence ($\gamma = 95\,\%$), to prove a controllability of 90 %, 29 test persons are needed without a single case of uncontrollability occurring. Formula 7.1 describes the criterion for the number of test subjects needed, as a function of the level of confidence and the proportion of controllability required for approval.

$$n = \frac{\log_{10}(1-\gamma)}{\log_{10}(\mu_x)} \tag{7.1}$$

n : Number of test subjects

γ : Level of Confidence

μ_x : Required proportion of controllability

In the "Code of Practice" (PReVENT 2009, p. 15) for the approval of C1, the effort for a statistical proof by tests with subjects is stated as too high. Again using the explained approach, at this level 299 test persons will be needed without accepting one negative test result. If negative results occur, the number of tests needed increases. As the minimum number of test persons is defined by the 5 % limit (see Formula 7.1), the rate of occurrence of this specific event (in this case, zero uncontrollable tests) is 5 % as well. This "success probability" of controllability testing, describing the likelihood that a controllability level can be proven with a specific number of test subjects depending on the expected controllability within the collective of drivers, can be calculated (see Table 7.2).

For low controllability proportions the success rate is low, even at a higher numbers of tests. Vice versa, for the example described in the "Code of Practice" (PReVENT 2009, p. 15), even if the controllability in the collective of drivers is on level C1, the testing will only have a 75 % chance of success. In most cases, if no transferability

Table 7.2 Number of tests needed versus success rate dependent on controllability proportions (Weitzel and Winner 2012, p. 19)

Controllability Level to be approved: ≥90 %		Success Probability			
Uncontrollable Events (k)	Minimal Number of Tests Needed (n), (Error Probability of 5 %, Rounded)	Expected Controllability Proportion in the Tested Group			
		90 %	95 %	97 %	99 %
0	29	4.7	22.6	41.3	74.7
1	46	4.8	32.3	59.7	92.3
2	61	4.9	40.6	72.3	97.7
3	76	4.7	46.9	80.1	99.3
4	89	4.9	53.9	87.0	99.8
5	103	4.8	58.9	91.0	99.9

can be proven such an effort is unjustifiable, especially because it has to be repeated for each system specification. Therefore, for the separation of classes C1 and C2 expert judgment is a commonly used method (Fach et al. 2010, p. 429).

7.4 Controllability Assessment for Unintended Reaction Scenarios

Apart from difficulties for the objective assessment of controllability on a higher level than C2, the focus of the ISO 26262 shows another challenge in its application to ADAS functions. The standard is intended for the functional safety of electric and electronic systems for vehicles up to 3.5 t. But hazardous situations caused by ADAS functions are not only a result of hard- or software failures. As they rely on environmental sensor systems, unintended reaction scenarios can be a result of incomplete information about the environment, e.g. due to limited capabilities or the number of available sensors, or also a misinterpretation of the situation by either the driver or the system. In these cases the system works within its specification but nevertheless the reaction is unintended. In addition, these "failures" are difficult to detect and additional sensors do not necessarily solve this problem, as they relocate the problem to the question of which sensor should be trusted—a strategy that increases costs rapidly. The expectable rates of misdetection and misinterpretation of a system are closely connected to the situation and to the utilization profile and cannot be determined in the same way as for hardware components.

This issue is not clearly addressed by ISO 26262. However, as the effects for the driver or other involved persons are considered to be equal, the methodology and testing described in the standard and according to the Code of Practice (PReVENT 2009, p. 15) should be feasible for these questions. Nevertheless, the transferability of the ASIL to this problem, in the meaning of absolute failure rates, is questioned in

		C1	C2	C3
S1	E1	QM	QM	QM
	E2	QM	QM	QM
	E3	QM	QM	A
	E4	QM	A	B
S2	E1	QM	QM	QM
	E2	QM	QM	A
	E3	QM	A	B
	E4	A	B	C
S3	E1	QM	QM	A
	E2	QM	A	B
	E3	A	B	C
	E4	B	C	D

Fig. 7.1 ASIL determination matrix (ISO 26262 2009, Part 3, p. 10)

some cases (Ebel et al. 2010, p. 396). The appropriate limiting risk measure depends on the system boundaries chosen. If the system for itself is taken as a function within the car, technically motivated levels like in ASIL should be useful. If the system boundaries consider the vehicle within its environment, the limiting risk is more to be seen in comparison to the driver capabilities.

7.5 Analysis of Unintended Reaction of ADAS

In order to be able to identify the minimum number of test scenarios needed to prove relevance for the field, the unintended reaction is examined and categorized. In parallel, the driving situation is analyzed in order to identify characterizing situational factors. These situational factors are then discussed and evaluated in relation to the categories of the unintended reaction. In combination with the anticipated probability of occurrence of the situational factor, the overall relevance of the factor is determined. Thereby a ranking of the situational factors on the unintended reaction scenario should be possible. Based on this, the minimum set of needed test situations can be identified. In the following these test situations are referred to as "Necessary Test Cases". The approach is to start at a "Best Case" of controllability for the system and add situational elements which potentially diminish the controllability. In combination with the probability of occurrence, their relevance could be identified similar to the ASIL Matrix in ISO 26262 (2009, Part 3, p. 10) (see Fig. 7.1), even though, as discussed earlier, the absolute levels are not transferable.

For the definition of the causes of unintended reactions of an ADAS function, two approaches are possible. The first approach assumes that the reaction is unintended by the driver and contradicts the planned maneuver or anticipated behavior of the vehicle

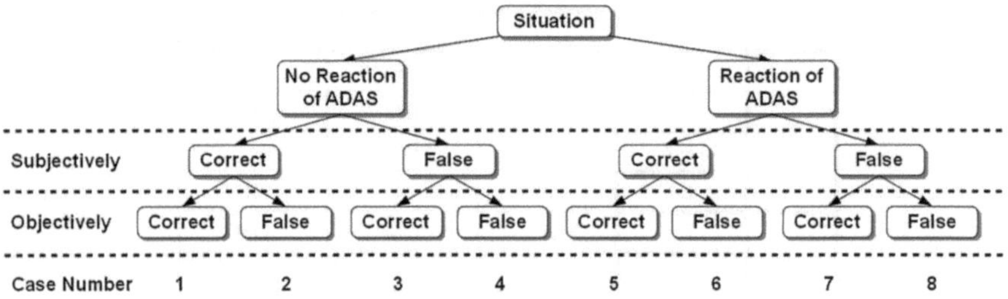

Fig. 7.2 Characterization tree for unintended reaction of ADAS

or the system. The second approach is based on an objective situation assessment, choosing the best available option within the situation and comparing the actual reaction of the system to that "best option". However this "best option" is not clearly definable in many situations, especially if it is necessary to anticipate the situation development including the behavior of other traffic participants. Even an "after-the-fact" evaluation, where the situation development is completely known, can be difficult, as the other traffic participants may have reacted differently if the reaction of the analyzed vehicle had been different. The two approaches are combined to an unintended reaction characterization tree as shown in Fig. 7.2.

As the aim of the test situation is to reveal the controllability of the unintended reaction, the situation should be as distinct as possible for the test person. Therefore, the most appropriate cases out of the characterization tree need to be identified. It has to be considered that after the incident, the driver will try to build an internal model of the system functions that match the experienced behavior (König 2012, p. 36). A contradiction between the driver's point of view and the "after-the-fact" evaluation of the situation, therefore, will cause bias, as the driver's reaction within the situation could be delayed due to additional decisions needed and an inconsistent internal model. To avoid this, in the definition of the Necessary Test Cases, both approaches should lead to the same subjective and objective conclusion. Assuming that only false reactions are critical, just the cases number 4 and 8 are suitable for controllability assessment.

The technical causes of unintended reaction are highly dependent on the sensor system, its signal processing, criticality estimation algorithm and rules for decisions about actions. The analytical identification of causes dependent on situational factors will lead too far into the field of sensor technology and post processing and will not be addressed here. An exception is, if the driver is planning a maneuver in the near future, e.g. an overtaking or lane change. In this case, as long as ADAS are not able to detect the driver's intentions, it is supposed that an unintended reaction is more likely.

To get a real and relevant situation where the driver is urged to intervene, a potential hazard must be perceivable (Muttart 2005, p. 3). Following the argumentation of ISO 26262, this hazard is a pending crash. Considered are two types of crashes, colliding with objects, including leaving the road, or colliding with other vehicles/other traffic participants. In the second case, the likelihood of hazardous objects depends on the traffic density. The objects are moving and controlled by a human being and may be able to avoid the accident by their own actions. So their ability of controlling the situation needs to be taken into account as well.

In summary, to compose a controllability situation, the cause must be distinct for the concerned person and the hazard must be present and urgent to trigger a reaction. This urgency is due to the limited time for reaction. Also, enough time for reaction must be available to enable the person to maintain control. This narrows the time span where controllability could be observed. This controllability time span is limited by two values, the last distance where a collision is just avoidable (at this point the time-to-react (TTR) is zero) (Hillenbrand 2008, p. 112ff), and the maximum distance where the situation can be perceived as critical by the driver. In addition, the reaction's dependency on internal/mental information processing and driver capabilities and character induces variation. Information processing in situations of unfamiliar vehicle behavior cannot be explicitly divided into phases with distinct values or described as a combination of these phases, because the processes can be serial and/or parallel and are not independent of each other (Olson and Farber 2003, p. 321). A suitable and valid driver model for the described problem would be the solution, but is not known. The existing driver models are able to address the typical driver behavior, but for controllability figures of 90 or 99 % the problem of the high numbers (see Table 7.2) occurs again and obviates any validation. As the individual processes cannot be observed in isolation, the minimum observation time span and other mental processes possibly involved have to be determined. However, the test results of naïve persons will always show scatter, sometimes with very broad variances. To achieve transferability, the source of the scatter and offset in the experiments must be analyzed. A differentiation must be made between scatter due to variation of driver behavior in steady situational conditions and scatter and offset due to variation of situation. This allows a prediction about the impact of situational variations and thereby simplifies the selection of relevant test cases. To enable such analysis, the assessment criterion for controllability has to be defined.

7.6 Assessment Criterion for Controllability

In the Code of Practice (PReVENT 2009, p. A50) the criterion for controllability is a nominal scale value. It differs according to whether or not the crash is avoided by the reaction of the considered person (Fach et al. 2010, p. 431). This binary assessment criterion is needed in the course of the risk level determination. In general, risk is determined as the probability of occurrence of a given hazard and its severity (ISO 31000 2009, p. 8 and 13). The assessment is based on the probability that the virtual hazard is transferred to a real accident. For testing with naïve persons the option of a real accident is not available. At the same time, the situation must be perceived as threatening by the person to provoke relevant reactions (Muttart 2005, p. 3 and 11). In some cases, deformable targets are used which are very close to the real accident but still with some trade-offs in real appearance. As discussed in the last section, to assess controllability the reaction must be observed to the point where no time to react is left. Subsequent reactions will only reduce the severity of the crash, which is assumed to be determinable by calculation. In summary, a real contact is not necessarily needed for the controllability assessment, as long as the last possible distance of reaction, called the point-of-no-return is included in the assessment period. If the driver reacts within the period before the point-of-no-return, the next question is whether the reaction was appropriate and intense enough to avoid the collision. Defining the point-of-no-return as the last possible moment for starting a lateral or longitudinal intervention means that the intensity of the counteraction needed for intervention increases from the start of the situation to that point where it reaches its maximum, limited by the maximum longitudinal or lateral force available. Time lags due to response characteristics of the system have to be added to this.

7.7 Situational Analysis

Theoretically, many factors influence a driving situation. Moreover, the possible detailing of these factors is nearly unlimited. A common approach is to classify these parameters in three parts (König 2012, p. 34):

- **Environment:** e.g. other traffic participants, weather, lighting, road condition…
- **Driver:** e.g. driving capabilities, internal model, attention, fatigue, character…
- **Vehicle:** e.g. vehicle behavior, speed, acceleration…

Many of these factors have interdependencies, for example all vehicles and drivers are influenced by the weather and therefore their driving behavior could change. For the controllability assessment, choosing many situational parameters and detailing the situation will cause a reduction of the probability of occurrence and thereby reduce the impact of the specific situation on the risk assessment. As a starting point, the unintended reaction is divided into three elements: causes, hazards, and reaction delaying factors. The influences of situational parameters in these three elements have to be discussed. To allow this, a start set of parameters for a generic situation definition is used and discussed in terms of impact on controllability. They are divided into two parts, parameters characterizing the state of the driver and environmental parameters describing the situation of the vehicle(s) and driver.

7.7.1 Driver Parameters

The driver influence factors according to Kopf (2005, p. 119) can be divided into three classes, depending on the rate of change. The slowest changing factors change over months or years, e.g. driving experience, learning effects or character of driver. For the present purpose they are not considered as explicit factors, but are taken into account by choosing an appropriate driver collective. The next class includes factors changing within a day or hours, e.g. drowsiness, drugs/alcohol. ISO 26262 (2009, Part 3, p. 9) defines that the driver should be in good constitution, therefore, influences due to these factors are also not considered. The last class comprises factors that change within minutes or seconds and therefore are relevant for the described tests because these factors can change within the test situation and thereby directly change the driver's behavior. According to Kopf (2005, p. 119), these factors are:

- **Driver intention**
- **Situation awareness:** concluded from cognition, perception and anticipation
- **Looking away:** as a conscious decision
- **Visual distraction:** due to a prominent stimulus outside the focus
- **Vigilance/Attention**
- **Strain/Stress**

From the above list, the driver intention is outstanding as it comprises the maneuvers planned in the near future and thereby, for example, influences the viewing direction (like to mirrors).

Following the best case approach mentioned above and the analysis of the unintended reaction, the consequences of variation in these parameters are estimated or assessed based on existing studies and data. This results in a ranking of situational factors that are expected to change the controllability.

To increase transferability and reduce test effort, driver models are very popular to simulate the driver behavior in different situations. In specific cases of driving dynamics, the driver is characterized as a transfer element (Mitschke and Wallentowitz 2003, p. 642 ff) in control theory. For unintended reaction scenarios, including warning elements, the information perception and processing of the driver is expected to consist of parallel and serial elements that are superimposed and interact with each other. These elements cannot be observed in isolation and feasible driver models are not known. However, if relevant driver parameters for unintended reaction scenarios can be identified and measured with appropriate accuracy, conclusions about the requirements for a driver model will be possible.

7.7.2 Environmental parameters

Different publications deal with the generic definition of potential driving situations (Domsch and Negele 2008, p. 7; Fastenmeier 1995, p. 48ff), or catalogues of situations. Reichart (2001, p. 52) focuses on situational parameters describing the road and environmental conditions which can be considered to be analogue to each other. After adaptation and simplification to match the requirements of the controllability assessment, the factors used are summarized in Table 7.3.

For the risk assessment of ADAS, the level of detailing of these influence parameters is crucial. By detailing the influence parameters, the overall rate of occurrence for the then more specific situation is lower. Assuming that the safety level needed for the system is derived from the highest risk of all situations, detailing will decrease the

Table 7.3 Environmental factors

Environmental factors	Road conditions
Light intensity	Number of lanes per direction
Visibility	Type of lane separation
Road friction coefficient	Lane width
Traffic density	Slope
	Banking
	Curvature

required safety level. A strategy to cope with this challenge is needed. It must take into account the detailing level of the different factors influencing the assessment of the controllability and allow a quantification of the coverage of situations at the different detailing levels. Although these requirements seem to be demanding, they must be fulfilled to enable a relevant and objective controllability assessment.

7.8 Conclusion and Outlook

For the assessment of the global characteristics of ADAS a more global/inclusive approach is needed which takes into account the whole system of traffic situation, vehicles and the driver and other involved persons. For the assessment of controllability, another challenge is added, as there is no predefined "use-case" for the controllability of unintended reactions of ADAS.

In addition, it is not economically feasible to address all possible unintended reaction scenarios. Therefore, it is necessary to identify a minimum number of relevant and reliable test scenarios for approval. The approach described here has been developed to this end. The unintended reaction has been analyzed and matched with factors for a generic situation definition. Based on this, the definition of categories of situational factors was proposed. This will allow the definition of necessary test cases and a rating of their relevance. The next steps are to carry out a detailed matching of the situational parameters with the elements of the unintended reaction and examine the feasibility of concrete relevance factors. The achievable accuracy for the factors has then to be analyzed. In addition, criteria for controllability need to be developed to enable a more detailed and transferable assessment. If this approach is applicable and successful, it can reduce the effort for approval testing. If it is not suitable, the reasons have to be discussed and the implications for the development of ADAS have to be concluded.

References

Domsch, C., Negele, H.: Einsatz von Referenzfahrsituationen bei der Entwicklung von Fahrerassistenzsystemen; 3. Tagung Aktive Sicherheit durch Fahrerassistenz, April 07–08, Garching, Germany (2008)

Ebel, S., Wilhelm, U., Grimm, A., Sailer, U.: Ganzheitliche Absicherung von Fahrerassistenzsystemen in Anlehnung an ISO 26262; Integrierte Sicherheit und Fahrerassistenzsysteme 26. VDI/VW-Gemeinschaftstagung, October 06–07, Wolfsburg, Germany (2010)

Fach, M., Baumann, F., Breuer, J.: Bewertung der Beherrschbarkeit von Aktiven Sicherheits- und Fahrerassistenzsystemen an den Funktionsgrenzen; Integrierte Sicherheit und Fahrerassistenzsysteme 26. VDI/VW-Gemeinschaftstagung, October 06–07, Wolfsburg, Germany (2010)

Fastenmeier, W.: Autofahrer und Verkehrssituation - Neue Wege zur Bewertung von Sicherheit und Zuverlässigkeit moderner Straßenverkehrssysteme, Mensch-Fahrzeug-Umwelt, Bd. 33, Verlag TÜV Rheinland Köln (1995)

Hillenbrand, J.: Fahrerassistenz zur Kollisionsvermeidung, Reihe 12, Verkehrstechnik, Fahrzeugtechnik, No. 669, VDI-Verlag Düsseldorf (2008)

Hoffmann, J.: Das Darmstädter Verfahren (EVITA) zum Testen und Bewerten von Frontalkollisionsgegenmaßnahmen, Reihe 12, Verkehrstechnik, Fahrzeugtechnik, No. 693, VDI-Verlag Düsseldorf (2008)

ISO 31000: 13.11.2009, risk management—principles and guidelines (2009)

ISO DIS 26262: 06.2009, Road vehicles— functional safety (2009)

König, W.: Nutzergerechte Entwicklung der Mensch-Maschine-Interaktion von Fahrerassistenzsystemen. In: Winner, H., Hakuli, S., Wolf, G. (eds.) Handbuch Fahrerassistenzsysteme. Vieweg + Teubner Verlag, Wiesbaden, pp. 33–42 (2012)

Kopf, M.: Was nützt es dem Fahrer, wenn Fahrerinformations- und -assistenzsysteme etwas über ihn wissen? In: Maurer, M., Stiller, C. (eds.) Fahrerassistenzsysteme mit maschineller Wahrnehmung. Springer Verlag, Berlin, pp. 117–140 (2005)

Mitschke, M., Wallentowitz, H.: Dynamik der Kraftfahrzeuge. Springer Verlag, Berlin (2003)

Muttart, J.W.: Factors that influence drivers' response choice decisions in video recorded crashes. SAE Technical Paper 2005-01-0426 (2005)

Olson, P., Farber, E.: Forensic Aspects of Driver Perception and Response, 2nd edn. Lawyers & Judges Publishing Company Inc., Tucson (2003)

PReVENT: Code of practice for the design and evaluation of ADAS, 13.08.2009 (2009)

Reichart, G.: Menschliche Zuverlässigkeit beim Führen von Kraftfahrzeugen, Reihe 22, Mensch-Maschine-Systeme, No. 7, VDI Verlag VDI-Verlag Düsseldorf (2001)

Weitzel, A., Winner, H.: Ansatz zur Kontrollierbarkeitsbewertung von Fahrerassistenzsystemen vor dem Hintergrund der ISO 26262, 8. Workshop Fahrerassistenzsysteme, September 26–28, Walting, Germany (2012)

Chapter 8
Design and Safety Analysis of a Drive-by-Wire Vehicle

Peter Bergmiller

8.1 Increasing Complexity in Modern Vehicles

In the Federal Republic of Germany, more than 700,000 people were employed in the automotive industry in 2010. According to the German Federal Transport Authority (Kraftfahrt-Bundesamt), the industry branch generated an overall turnover of about 315 Billion Euros. Therewith, the automotive industry contributes hugely to the national output of Germany and thus is vital for the further economic success of the country (Legler et al. 2009). At the same time, the complexity of modern vehicles is continuously increasing and vehicle development is becoming more and more challenging. Additional and more complex functionalities from different domains (ergonomics, entertainment, etc.) are demanded by the customer (Sangiovanni-Vincentelli 2007). Adaptations and extensions of the electrics and electronics (EE) of the vehicle, especially the integration and interaction of previously independent functions, significantly contribute to meeting these demands (Arbitmann et al. 2011; Sinha 2011; Pruckner et al. 2012). Thus, the number of interconnections and interdependencies within the EE system rapidly increases on functional and hardware level (Schäuffele and Zurawka 2004). This furthermore pushes complexity.

In parallel, operational safety of modern vehicles has to be maintained, or better, improved. This generates strongly conflicting goals between additional functionalities and proof of sufficiently safe operation of these increasingly complex systems. With electric vehicles joining the market, meeting both targets becomes even more challenging. Depending on the drive train structure, electric vehicles can provide powerful means to intervene into vehicle handling, e.g., due to torque vectoring.[1]

[1] Torque vectoring refers to an approach where individual wheels of a vehicle are driven with individual drive torques. When driving the wheels at one side of the vehicle with a different torque than the wheels of the opposing side, an additional yaw moment is generated. For further information including evaluation of safety criticality see, e.g., Euchler et al. (2010).

P. Bergmiller
Institute of Control Engineering,Technische Universität Branuschweig, Hans-Sommer-Str. 66, D-38106 Braunschweig, Germany
e-mail: p.bergmiller@tu-bs.de

M. Maurer and H. Winner (eds.), *Automotive Systems Engineering*, DOI: 10.1007/978-3-642-36455-6_8, © Springer-Verlag Berlin Heidelberg 2013

These means are controlled by the EE system, and thus a failure of the EE system can result in fatal crashes that are hardly avoidable even for a skilled driver. Consequently, profound further development in the field of vehicle electronics is crucial not only for the German automotive industry to maintain its leading global position based on innovative functionalities, but also to make driving safer. This necessity is strongly supported by a technical survey issued and funded by the Federal Ministry of Economics and Technology in Germany (BMWi) (Bernard et al. 2010). The survey especially demands fundamental reconsideration of the established electronics architecture in series vehicles in terms of performance, reduction of complexity and functional safety. This is regarded as a key factor for further technical progress in the field of vehicle electronics and economic success. Turning from the classical perception of the vehicle as a number of interconnected parts to a more holistic functional perception of the overall vehicle can be a way to achieve this goal. This approach is followed at the Institute of Control Engineering at TU Braunschweig (Maurer 2010) and also slowly starting to be adopted by companies (Abele 2012; Papadopoulos et al. 2001).

To support this development, this contribution introduces (a) a flexible experimental vehicle (MOBILE) for fundamental investigations in the field of vehicle electronics that is built up at TU Braunschweig; and (b) an hierarchical approach for focused evaluation of functional safety of vehicles with a high degree of functional redundancy, functional integration, and complexity due to by-wire control.[2] For MOBILE, especially the functional and hardware/software architecture of the drive-by-wire system are introduced. The applicability of the hierarchical approach is demonstrated by analyzing MOBILE in terms of functional safety. Thereby, a simplified hazard analysis according to ISO 26262[3] that is based on expected use cases of MOBILE delivers the safety goals[4] that have to be met by the design of the drive-by-wire system.

8.2 The Experimental Vehicle MOBILE

The experimental vehicle MOBILE is custom built by the Institute of Control Engineering and the Institute of Engineering Design at TU Braunschweig. The intended purpose of the vehicle is to serve as a tool for a variety of future research projects on vehicle dynamics and mechanical or electric/electronic components. Still, the vehicle itself is also subject to research activities. The highly safety critical drive-by-wire system in combination with a high degree of functional integration require novel approaches in system design. Resulting, costs for hardware units in the vehicle can

---

[2] In this context, "by-wire" control means that actuators in the vehicle are controlled purely electronically without any mechanical or hydraulic linkage between the actuator and the driver. MOBILE implements by-wire control for braking, steering, and the drive motors.

[3] ISO 26262: Road Vehicles—Functional Safety, edition 2011.

[4] According to ISO 26262, a safety goal is a "top level safety requirement as a result of the hazard analysis and risk assessment" (ISO 26262-1:2011, p. 14).

be reduced, and benefit for the driver is increased due to synergies between different functionalities. This contribution presents the basic actuator set-up of MOBILE and details the architecture of the part of the EE system needed for vehicle control. Other parts as the extended Human-Machine-Interface (displays, inputs other than brake and gas pedal), the battery management, the cooling system, and the knowledge management are neglected. These aspects are investigated and implemented in the MOBILE project but are not covered in this contribution. Accordingly, the following sections introduce the architecture[5] of the vehicle control function stepwise: Starting from general requirements, the architecture is detailed on different hierarchical layers and from different views.[6] Figure 8.1 outlines the steps for definition of the architecture. Thereby, requirements and constraints (step 1 in Fig. 8.1) guide the derivation of the basic mechanical and actuator set-up of the vehicle (step 2 in Fig. 8.1). Following, the steps 3 to 5 in Fig. 8.1 iteratively derive the functional/software and hardware architecture of the EE system on all hierarchical layers. Functional and software architecture partially merge at higher hierarchical layers. Thus, no dedicated software views will be presented on the individual layers. Figure 8.2 introduces the referenced hierarchical layers.[7] The classification into result layer, detailed layers, and assumed inputs given in the figure will be explained in Sect. 8.3. Step 6 in Fig. 8.1 briefly introduces some additional aspects from a software view. Finally, the overall system is evaluated with regard to goal achievement (Step 7 in Fig. 8.1).

8.2.1 Requirements, Constraints, and Principles

The definition of requirements constitutes the first step of product development. Analogously, some core-requirements (Table 8.1) and constraints (Table 8.2) guided the development of MOBILE. Additionally, further influences impact decisions on architecture and system development. In general, Maier and Rechtin (2009) distinguish "normative (solution based), rational (method based), participative (stakeholder based) and heuristic (lessons-learned)" (Maier and Rechtin 2009, p. 1) methodologies. To cover important normative and heuristic influences, Sect. 8.2.3 provides a summarized state-of-the-art on structures of drive-by-wire systems, used mechanisms, and best-practices. In Sect. 8.3, a novel method to analyze system safety

---

[5] ISO/IEC 42010:2007 defines architecture as follows: "The fundamental organization of a system, embodied in its components, their relationships to each other and the environment, and the principles governing its design and evolution".

[6] The architecture of a system can be described from different views depending on the goal of the description. Examples can be business views, process views, but also functional or hardware views (Masak 2010). For some of these views guidelines for standardized diagrams exist, e.g., UML for software systems (Starke 2008).

[7] In the following, hierarchical layers are always referred to as "X level" or "X layer" (with X standing for vehicle, system, etc.) to clearly distinguish between referencing of a layer and the general use of the words system, component, or element to refer to certain elements independent from layers.

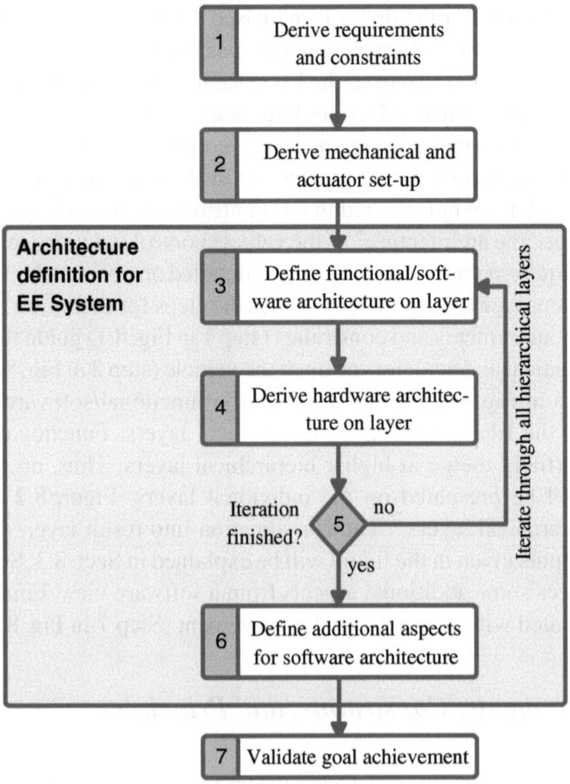

Fig. 8.1 Architecture definition process for MOBILE

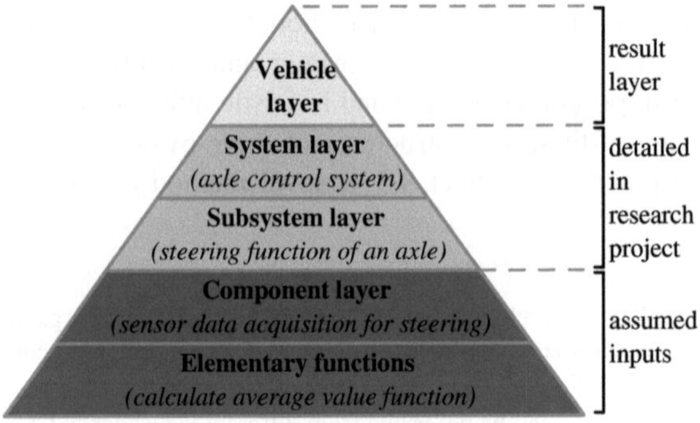

Fig. 8.2 Hierarchical layers for architecture definition

Table 8.1 List of requirements

Providing a flexible experimental vehicle	
Requ. 1	The vehicle has to feature a *high degree of mechanical and electric/electronic modularity* that allows easy exchange of hardware components for research and testing. At the same time, the base configuration of hardware and software of the vehicle has to be powerful to support various and highly dynamic driving experiments.
Requ. 2	The software running on the electronic control units shall be easily accessible and exchangeable. The vehicle can be seen as an "*open-source vehicle*" that is readily available for any research tasks. Compatibility with the graphical programming environment Simulink of Mathworks is desired to support code re-use and shorten training periods.
Guaranteeing sufficiently safe testing	
Requ. 3	Although the experimental vehicle is only operated on test tracks, the vehicle should *fulfill basic safety requirements* and tolerate one independent fault[a] with a given probability. Details will be given in Sect. 8.3.1.
Requ. 4	The degree of hardware redundancies for safety purposes shall be reduced. In turn, the safety concept shall *exploit functional redundancies* between different types of actuators that are available in the vehicle.

[a] A fault is an "abnormal condition that can cause an element or item to fail" (ISO26262-1:2011, p. 7).

Table 8.2 List of constraints

Const. 1	MOBILE is a university only project and thus benefits from graduate and undergraduate students writing their theses on individual development tasks. Accordingly, these work packages have to be clearly defined and proper documentation plays a huge role in the project.
Const. 2	All tasks worked on by the project partners have to stem from the according core fields of research, i.e., vehicle electronics or design of mechanical parts. Other parts have to be sourced externally, e.g., actuators.
Const. 3	The project is subject to strict financial limits. Resulting, mostly "off-the-shelf" components have to be relied on.

is presented (rational methodology). Participative aspects are not detailed in this contribution.

8.2.2 Mechanical and Actuator Set-Up of MOBILE

The mechanical and actuator set-up is especially driven by the requirements on modularity and universal applicability of the vehicle *(Requ. 1)*. Accordingly, MOBILE was designed as a full electric vehicle with by-wire control for propulsion, braking, and steering:

The *electric drive concept* contributes to a powerful base configuration and ensures flexibility in the longitudinal behavior of the vehicle. The research project

InDrive demonstrated that the longitudinal behavior of a target vehicle can be simulated by a powerful carrier vehicle given little latencies in traction control (Cornelsen et al. 2011). The electric drive concept of MOBILE with a peak power of about 100 kW per wheel and 400 kW total can fulfill these requirements. Additionally, independent driving of each wheel allows yaw control via torque vectoring. The benefits and risks of such systems for vehicle handling are, e.g., evaluated by Euchler et al. (2010); Piyabongkarn et al. (2007), or Rohe (2012). Furthermore, the electric components in combination with by-wire control enable good modularity. Combined with appropriate mechanical design, drive units for an axle can be removed or replaced easily.

Four-wheel steering furthermore extends the fields of application of the vehicle. In general, the steering system can be implemented as a rack actuating type, a tie-rod actuating type or a knuckle actuating type (Park et al. 2005). In order to be able to individually steer each wheel and based on the components available on the market, the tie-rod actuating type was implemented. Thus, different steering geometries and steering concepts can be emulated by simple software modifications. In terms of performance, Wilwert et al. (2005) consider a ±40 degree steering angle per wheel as desirable for a drive-by-wire system at a steering rate of up to 40 degrees per second (see also Heiner and Thurner (1998) from the view of an OEM[8]). This steering angle approximates typical characteristics of steering systems in non-by-wire series vehicles with front wheel steering (Pfeffer and Harrer 2011). For MOBILE, each individual steering system features an adjusting range of approx. ±43 degrees and a steering rate of 130 degrees per second at nominal load. Thus, also highly dynamic maneuvers are possible.

The electro-mechanical braking system of MOBILE is designed to outperform most hydraulic brake systems in terms of reaction times. This allows precise slip control and research towards seamless integration of recuperative and mechanical braking for optimized recuperation (Pruckner et al. 2012). Additionally, the braking system renders hydraulic components in the vehicle unnecessary and thus does little impact vehicle package. The individual brake units at each wheel were designed by Vienna Engineering to ensure a 1 g deceleration of MOBILE at a maximal weight of 2.100 kg including passengers. First tests with the braking system on an experimental set-up indicate that the brake system will outperform these requirements. The safe state of each individual brake in the project MOBILE is defined as a state without any brake torque as also preferred in literature (Johannessen et al. 2004; Sinha 2011).

The vehicle is powered by a *modular power supply* consisting of two independent units providing 300, 48, and 12 V each. Currently, the main source of power of each unit is based on lead-acid batteries resulting in the given voltage level of approximately 300 V. In future, this battery pack is planed to be exchanged by lithium-ion batteries with a pack voltage of 400 V and higher energy density. 48 V and 12 V are mandatory to supply the externally sourced actuators and vehicle electronics (48 V for steering, 12 V for braking and vehicle electronics, *Const. 3*). The low voltage circuits are supplied by the main battery pack via DC-DC converters and buffer batteries

[8] Original Equipment Manufacturer, e.g., BMW, AUDI, Toyota for the automotive industry

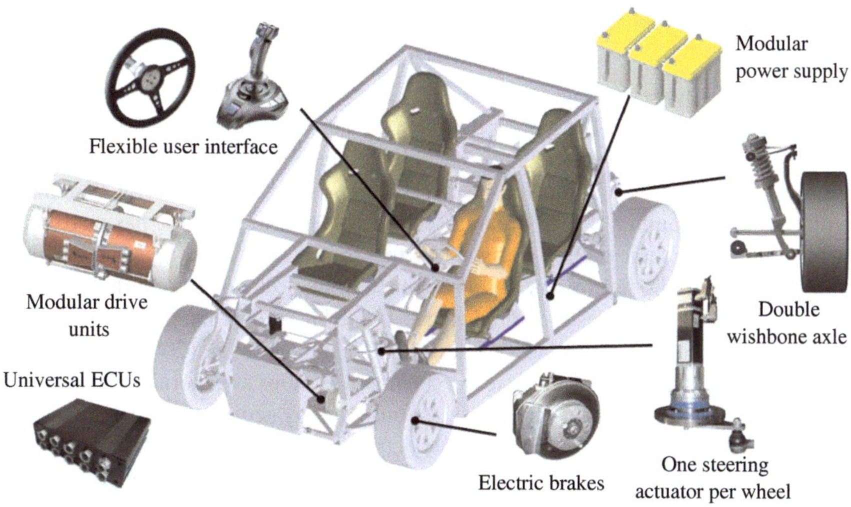

Fig. 8.3 Actuator equipment and mechanical set-up of MOBILE; ECU: Electronic Control Unit

with low capacity. Two independent power supply units reduce the overall failure[9] rate (*Requ. 3*) and limit the required currents per battery pack at peak load. Steering and braking actuators at diagonal positions in the vehicle are connected to a common power supply. Resulting, vehicle handling is less effected in case of failure of one power supply. This corresponds to the design of braking systems in series vehicles (ECE R13[10]) and is frequently replicated for brake-by-wire systems proposed in literature (Rieth 2012; Papadopoulos et al. 2001). The powerless steering actuators are back-drivable and thus allow to be moved by torque at the wheels applied by the drive motors given a suitable axle geometry (Dominguez-garcia et al. 2004).

The by-wire architecture allows flexible design of the *user interface*. All input devices can be exchanged on demand. In a base configuration, a force-feedback steering wheel, a force-feedback brake pedal, and a gas pedal are available to the driver. Also, these units provide feedback on the road surface and the current driving condition. A flexible touch-screen based visualization allows easy access to all measurements taken in the vehicle (Bergmiller et al. 2011a).

To conclude the introduction of the mechanical and actuator set-up of MOBILE (step 2 in Fig. 8.1), Fig. 8.3 summarizes the equipment of MOBILE and provides an overview of the mechanical set-up. Further details on the components can be found in Bergmiller and Maurer (2012). Summarized, the actuator set-up facilitates high flexibility of the vehicle. At the same time, it provides a high degree of functional redundancy between different types of actuators, which can be exploited by novel approaches to achieve functional safety.

[9] ISO 26262 defines failure as the "termination of the ability of an element to perform a function as required" (ISO26262-1:2011, p. 7).

[10] United Nations Economic Commission for Europe: Brake System Homologation.

8.2.3 Related Work for EE Systems of Drive-by-Wire Vehicles

This section provides an overview of the state-of-the-art for the design of safety critical by-wire systems. The gathered information on system structures and common practices serve as important input for the architecture derivation of MOBILE in steps 3 to 6 given in Fig. 8.1. Figure 8.4 proposes a generic view on by-wire systems as perceived by the author. The following section will first explain the basic structure of the figure and then outline the state-of-the-art for individual key aspects. The numbers given in the figure serve as a reference in the following paragraphs.

❶ Most by-wire systems for vehicles investigated in research, e.g., by Armbruster (2009), Heiner and Thurner (1998), Sinha (2011), Wilwert et al. (2005) or Zuo et al. (2005), split the system in two physically separated sections as generalized in Fig. 8.4. One section contains the user interface and consequently acquires data from the user, the other section controls the main vehicle actuators. Depending on the investigated system, the actuators are either steering actuators, brake actuators, other actuators as, e.g., for control of vertical dynamics, or combinations of these. Accordingly, the user input devices change. Input devices can feature actuators to provide additional feedback on the road surface and the driving situation to the driver.

All components are controlled by ECUs that are mounted close to the relevant actuators or sensors. Other designs that wire all components directly to one central controller as presented by Park et al. (2005) or several other research vehicles with

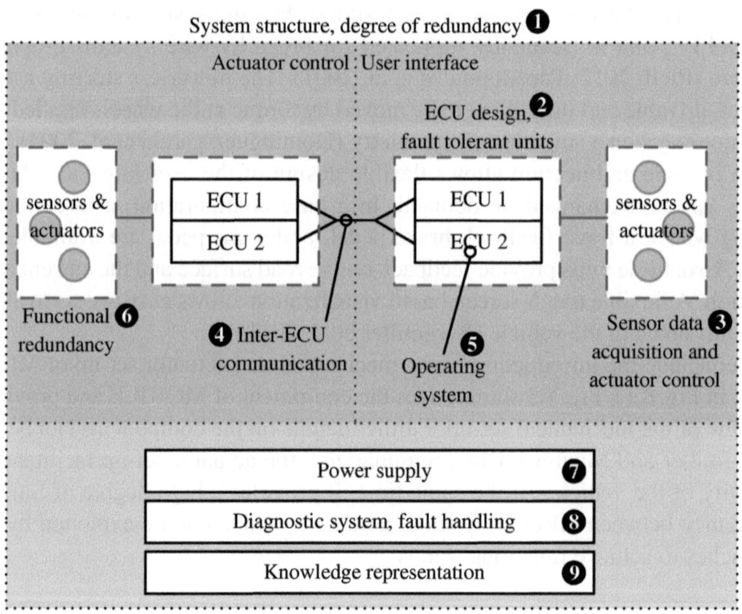

Fig. 8.4 Structure of a generic by-wire system as referenced in the state-of-the-art section; ECU: Electronic Control Unit

no focus on functional safety of the EE system are not further regarded for wiring effort, EMI,[11] and modularity reasons. Also, solutions with mechanical/hydraulic fall-back layer are not considered (Zuo et al. 2005). Each safety critical ECU is usually available redundantly as no single unit can—so far—achieve the required failure rates for automotive drive-by-wire systems. In general, the degree of hardware redundancy is kept as low as possible due to production costs. Two redundant components for one task in combination with a sufficiently powerful diagnostic and decision unit and fail-silent behavior of each component are assumed to be able to fulfill the required failure rates (Wilwert et al. 2005; Miller 2007). Several systems featuring this structure can be found in literature as, e.g., introduced by Hasan and Anwar (2008), Armbruster (2009), Sinha (2011), or Wilwert et al. (2005).[12] Such a combination of two (or more) ECUs is then regarded as fault-tolerant unit[13] ❷. If all units are permanently turned on, the system is denoted as operating in hot-standby mode (Neudörfer 2011). This shortens take-over times in case of failures of the primary unit as the secondary unit does not have to boot or initialize.

In some cases, redundant ECUs are replaced by a single ECU featuring a multi-core architecture and according board design in combination with special mechanisms to allow executing multiple safety critical functions independently on this platform. According strategies are, e.g., outlined in the RECOMP project (Motruk et al. 2012) funded by the Federal Ministry of Education and Research (BMBF) or by Philipps (2012). Amongst others, a dedicated software functions and storage management is required to proof independence of functions with regard to common cause failure.[14] Also on chip level, cost efficient means to detect and correct transient faults are investigated (Touloupis et al. 2005).

A further approach to reduction of hardware redundancies can be a network centric architecture, where nodes monitor each other mutually. In case of a failure of a single ECU, the other ECUs detect the failure and continue accordingly adapted operation. As a result, the number of ECUs can be reduced, but the complexity of each individual device increases (Kelling and Heck 2002; Johannessen et al. 2002; Sakurai et al. 2008). Resulting effects on costs have to be evaluated for each specific system. Especially, brake-by-wire can benefit from network centric architectures, as several redundant actuators exist that can be equipped with independent ECUs. Also, the vehicle can still be decelerated with two or three brakes available. For the project MOBILE, a combination of a network centric approach and classical redundancies seems reasonable.

The redundancy strategy is extended for sensors and actuators ❸. On the sensor side, typically at least triple hardware redundancy is applied to acquire multiple

[11] Electromagnetic interference.

[12] Note: For fly-by-wire systems at least quadruple redundancy for military aircrafts and higher degrees of redundancy for civil aviation are required (Collinson 1999).

[13] In a fault tolerant unit, a defined number of faults does not lead to a failure of the overall unit, e.g.,Wilwert et al. (2005).

[14] A common cause failure is a "failure of two or more elements of an item resulting from a single specific event or root cause" (ISO26262-1:2011, p.3).

sensor signals and perform majority voting to determine faulty measurements as, e.g., demonstrated by Bertacchini et al. (2005). Other approaches rely on analytical redundancy that replaces one or two sensors by software algorithms that derive additional "virtual sensor data" or diagnostic residuals for the investigated signal by comparison with other sensor data available in the system (Anwar and Niu 2010; Gadda et al. 2007; He et al. 2010; Kim et al. 2010). Also, fault tolerance strategies are frequently implemented mechanically and electronically within the sensors or actuators (Dilger et al. 2004). Especially, model based diagnostic algorithms can assist identification and treatment of faults already within the actuators (Isermann and Beck 2011; Muenchhof et al. 2009). Resulting, the required number of physically separated redundant units can be reduced. For steering, mostly two redundant actuators are available at one axle (Muenchhof et al. 2009; Wilwert et al. 2005; Zhen et al. 2005; Zuo et al. 2005). For the braking system, each wheel features an individual actuator (Isermann et al. 2001; Papadopoulos et al. 2001; Reichel and Armbruster 2011). The feedback actuators at the user input devices are mostly also classified as safety critical. Still, the criticality is lower than the one of the actuators at the wheels. Depending on the investigated system, developers implement these actuators as single actuators (Anwar and Niu 2010), redundantly (Wilwert et al. 2005) or provide mechanical back-up feedback (Pruckner et al. 2012; Zuo et al. 2005).

The system design according to the so far outlined guidelines ensures proper operation of the sensors and actuators and supports safe execution of the application algorithms on computational platforms. Still, a main issue in by-wire Systems is the communication between physically separated ECUs ❹. The according data bus systems have to be available at least in single redundancy, including physical separation in wiring as, e.g., outlined by Wilwert et al. (2005). Some research projects also propose more than two physically independent channels, e.g., Sinha (2011) and Sundar and Plunkett (2006). Additionally, the overall network has to feature given and precise timings to ensure that lost or delayed messages can be detected and a maximal round trip time can be guaranteed (Heiner and Thurner 1998). Wilwert et al. (2005) derive a maximal acceptable end-to-end response time for driver inputs to the steering actuators of 17.6 ms.[15] Beyond this limit, the vehicle becomes unstable. In applications, exclusively time-triggered data bus systems are relied on, e.g., TTCAN (He et al. 2010), TTP/C (Papadopoulos et al. 2001; Blanc et al. 2009), or FlexRay (Sinha 2011; Sundar and Plunkett 2006; Waraus 2009). Some research projects also investigate the applicability of Ethernet in combination with a time triggered extension (Müller et al. 2011). These bus systems provide precise and deterministic communication timing at the price of less flexibility for spontaneous adaptations during the design process (Mishra and Naik 2005). Then, the challenge arises to synchronize the user application with the network timings in order to ensure defined latencies and synchronization throughout the network (Sundar and Plunkett 2006). Different operating systems ❺ support this task as a modified OSEK (Sakurai et al. 2008), OSEK Time, FTCom (Wilwert et al. 2005), or recently also AUTOSAR (Mitzlaff et al. 2010;

[15] For comparison: In aviation, sensors are sampled about 100 times per second which roughly equals the minimal demands in the automotive field (data for A320, Collinson (1999)).

Tucci-Piergiovanni et al. 2011). Still, the applicability of each operating system has to be confirmed individually depending on the required precision of timings and the available computational resources provided by the network nodes.

Assuming a proper interaction and operation of the individual components within the vehicle was established, functional redundancies ❻ between different actuators and especially different types of actuators (steering, drive, brake) to achieve the overall safety goals can be exploited. Thereby, hardly any research projects are available that exploit these redundancies for a proof of safety in accordance with ISO 26262, but several projects investigate possible functional redundancies and cross-couplings between the individual actuators from a view of vehicle dynamics. Thereby, simulation or experimental vehicles serve as test beds (Arbitmann et al. 2011; Dominguez-garcia et al. 2004; Euchler et al. 2010; Hayama et al. 2008; Johannessen et al. 2002; Reinold et al. 2010). Further contributions from research groups in the field of vehicle dynamics, e.g., the groups of Gerdes at Stanford University and Trächtler at Universität Paderborn, exploit the capabilities of highly flexible experimental vehicles to make driving itself safer. Thereby, safety of the drive-by-wire system is not focused. These research results serve as a guideline to what vehicle control algorithms are already available or can be expected to be available in the near future. Accordingly, the design of the EE system of MOBILE was influenced.

To ensure the operation of the overall system, a fault tolerant power supply unit ❼ is mandatory. Typically, redundant systems with mutual isolation are implemented (Abele 2008; Kelling and Heck 2002; Sieglin 2009). The GM vehicle Sequel implements triple redundancy, and additional means to reconfigure the power supply in case of failure (Sundar and Plunkett 2006). Such central reconfiguration units for power supply systems are useful for failure compensation and frequently relied on-(Armbruster et al. 2006; Sundar and Plunkett 2006). But, they may also turn out as a weak spot of an architecture due to their huge impact on the system in case of failure. Typically, the power supply provides 12V to steering and braking actuators, unless the vehicle weight requires higher voltage levels (≥ 42 V) to cover the increased power demands (Wilwert et al. 2005; Sundar and Plunkett 2006).

Based on the components introduced so far, the by-wire system is operable. It should be able to maintain at least degraded operation after any first fault as demanded according to the state-of-the-art (Armbruster et al. 2006; Pruckner et al. 2012) and typical demands for licensing of vehicles (ECE R13). Still, a powerful diagnostic system ❽ has to be provided in order to ensure the detection of faults and to provide the basis for appropriate fault handling. As given above, most components already provide means to diagnose proper operation. In combination with network overarching monitoring mechanisms for timings and interfaces, a huge data base on the current state of the vehicle is available. To derive according actions from this data, different approaches, mostly relying on heuristics and probabilistic interpretations, were developed (Bergmiller et al. 2011b; Isermann and Beck 2011; Muenchhof et al. 2009; Schwall and Gerdes 2002). A challenge for these algorithms are the short execution times that have to be guaranteed. Additionally, the behavior of the diagnostic and action derivation system has to be predictable which renders most machine learning approaches unsuitable. Typically, the vehicle is regarded as non self-healing.

Thus, repair of defective components is not performed online. In aviation, self-restart of components are taken into account (Schroer 2008) to improve system safety. This idea is also investigated theoretically for the automotive domain (Pimentel 2003) but hardly followed for safety critical systems in real vehicles. In the project MOBILE, the idea is adapted for real application in the experimental vehicle (Bergmiller et al. 2011b).

Finally, the information on the overall system acquired by the diagnostics system can be put in relation and integrated in a "knowledge base" ➒. This knowledge[16] includes relevant information on the current capabilities of the vehicle and according maneuvers that can be executed. For autonomous vehicles, such investigations have already been made by Maurer (2000) and Siedersberger (2003). With increasing capabilities of the vehicle, the part focusing on the ego vehicle should be further emphasized and extended. The challenge becomes even bigger when different modules can be combined flexibly within one vehicle. This is so far not targeted sufficiently by research projects but investigated in the MOBILE project. Therefore, a flexible ability based self representation is implemented.

Conclusion and differentiation of MOBILE: Multiple research groups are actively working in the drive-by-wire field, but hardly any group targets the overall system including steering, braking and the propulsion function from a functional safety view. Usually, only one system is investigated, and mostly these systems are implemented on a test-rack or in combination with simulators and not in real vehicles. Some research groups as the ones of Gerdes and Trächtler (Beal and Gerdes 2010; Gadda et al. 2007; Trächtler and Niewels 2006; Reinold et al. 2010) have built up vehicles with extended by-wire functionality. Still, the safety of the onboard EE system is not investigated in detail for these vehicles. These groups focus on vehicle dynamics. Research results in this field are taken as boundary conditions for identification of possible functional redundancies in MOBILE.

Johannessen (2001) presents a modified Sirius vehicle (*SIRIUS 2001*) with individual steering and braking of each wheel and time-triggered communication for a network centric approach to safety. To some extent, also cross compensations between actuators to control the vehicle, e.g., steering by differential braking, are considered. The vehicle is based on a conventional propulsion system, and the wheel brakes are hydraulic but controlled by electrical pumps to generate pressure. Also, an analysis of the vehicle failure rate is performed resulting in a failure rate of 5.74 EE−8 catastrophic failures per hour but neglecting the power supply system. In a follow up project (FAR project) a model vehicle with four wheel steering, individual braking and four wheel drive was built up Johannessen et al. (2004b). Some results from the Sirius vehicle can be adopted and taken as reference for MOBILE. Still, flexibility requirements for tooling and full consideration of all components of the vehicle including power supply and an propulsion system capable of torque vectoring will require adaptations and extensions for MOBILE. Especially, the strongly

[16] Knowledge denotes the "awareness, consciousness, or familiarity gained by experience or learning" (Collins 2010). In the project MOBILE, the "self-awareness" of the vehicle is considered. The "experience" is provided at design time based on experiments or statistics.

network centric approach reduces flexibility of the Sirius vehicle for usage as a development tool.

The European project *SPARC* (Armbruster 2009; Reichel and Armbruster 2011; Sieglin 2009) stands out by thoroughly investigating a full by-wire concept for application in different vehicle classes (trucks and cars). The project presents a full drive-by-wire architecture that is applied to different experimental vehicles. The vehicles feature one steerable axle and a brake-by-wire system. The by-wire system includes redundant actuators and a degradation approach to handle faults. Still, the system architecture requires the memory in the network to be available in quadruple redundancy as all nodes have to be able to perform all computational tasks. Compared to this project, MOBILE features higher flexibility due to the given actuator set-up. Also, MOBILE targets an even lower degree of hardware redundancy of actuators and controllers by exploiting the functional redundancies in the vehicle instead.

In general, the design of MOBILE focuses the usage as an experimental platform and serves as a proposal for future series vehicles given the further progress of research in the given fields. This clearly distinguishes MOBILE from approaches for current series vehicles. Summarizing, the design of MOBILE, as will be presented in the next section, exploits several principles outlined above: Any ECUs, actuators, and data bus connections for one task should be available at most twice. Lower degrees of redundancy should be implemented if functional redundancies can be exploited. Sensors are implemented in triple redundancy. Later on, analytical redundancy can easily replace existing sensors. Safety critical components should be kept independent from each other wherever possible such that the overall vehicle can tolerate one independent fault and guarantee an emergency run interval. Communication in the vehicle will be time triggered and allow precise synchronization of applications throughout the network. Force Feedback is not regarded as a highly safety critical function in MOBILE as a skilled test driver is driving the vehicle. Resulting, the according actuators are not implemented redundantly, but mechanical open-loop feedback will ensure controllability.

8.2.4 Hierarchically Structured Architecture Derivation

This section presents the architecture of the EE system of MOBILE. The architecture is intended as a template for by-wire vehicles with high need for flexibility and safety at low costs. As will be outlined at the end of this section, basic ideas could also be transfered for cost efficient proof of functional safety in series vehicles. As mentioned, the architecture is derived top-down along the hierarchical layers introduced in Fig. 8.2. On each layer, at first the functional architecture is presented, then a suitable hardware architecture is derived that allows to execute all needed functions and fulfills requirements on modularity and functional safety (steps 3–5 in Fig. 8.1). The requirements given in Sect. 8.2.1, the hardware set-up introduced in Sect. 8.2.2, and already existing research results (Sect. 8.2.3) guide the architecture derivation.

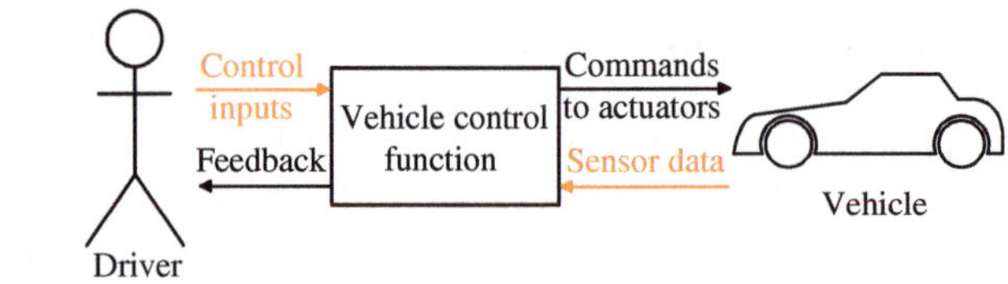

Fig. 8.5 Functional architecture of MOBILE on "vehicle layer"

8.2.4.1 Vehicle Layer

The fundamental functional architecture on "vehicle layer" is simple: It consists only of the vehicle control function if other comfort or add-on functions are neglected. The function acquires data from the driver (control inputs) and accordingly controls the actuators of the vehicle (commands to actuators). Vice versa, sensor data is gathered to execute the vehicle control function and to provide feedback to the driver. Figure 8.5 outlines these basic dependencies in the style of a UML context diagram. The hardware architecture is structured similarly to the functional set-up. It consists of the part of the EE system concerned with vehicle control. All other parts that are not relevant for the basic driving function are neglected. Individual hardware components are not distinguished at this layer.

8.2.4.2 System Layer

The "system layer" splits the vehicle control function into the most important components. Figure 8.6 shows the according functional view. Thereby, the special purpose of the vehicle as a development tool and the intention to exploit functional redundancies are already regarded:

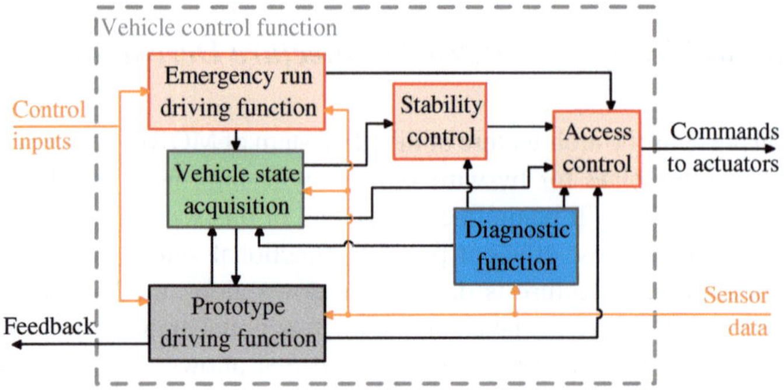

Fig. 8.6 Functional architecture of MOBILE on "system layer"

A *prototype driving function* that is realized by software/hardware under development controls the vehicle actuators in experimental mode. When implementing this prototype function, the developer should be provided full access to the vehicle including all sensor data and access to all actuators (*Requ. 1/2*). The prototype driving function allows to provide feedback to the driver via interface devices as the steering wheel or the brake pedal. The functionality implemented by the prototype driving function can hugely vary, e.g., four-wheel steering vs. steering of only one axle with adjustable steering ratio.

As a basis for the prototype driving function and to provide data to other units in the vehicle, the *vehicle state acquisition* function gathers all available sensor data related to vehicle dynamics and derives the current vehicle state. Thus, the vehicle state acquisition contributes significantly to the tooling character of the vehicle (*Requ. 1*). The vehicle state acquisition gathers its data throughout the network. Depending on the state of the according nodes, the vehicle state acquisition can rely on data from the emergency run or the prototype driving function.

Jointly, the prototype driving function and the vehicle state acquisition function allow the control of the vehicle. Further components are added to ensure safe driving. This includes an *emergency driving function*, a *stability control* module, and the *access control*. The emergency driving function provides basic control of actuators in a "fall-back" manner. Thereby, the actuators are operated in their most basic mode of operation with minimal usage of extra sensors. No cross-couplings or dependencies with other functions exist. Still, the emergency driving function provides the driver full access to steering actuators, brakes, and drive motors (*Requ. 3*).

As MOBILE is only equipped with one actuator for steering, braking, and drive at each wheel, functional redundancies have to be exploited for safety. Accordingly, the stability control system operates, on the one side, as a conventional stability control system known from series vehicles. On the other side, it compensates the failure of a single actuator by adapting its control strategy (*Requ. 4*). To detect each failure state, the stability control relies on the data provided by the *diagnostic system*.

The diagnostic system monitors the current state of the vehicle from an electric/electronic point of view. Faults within components and resulting possible failures are indicated to all nodes of the network.

Finally, the access control determines which driving function may control the actuators: the prototype driving function, the emergency run driving function or the stability control in case of actuator failures (*Requ. 3*). Thereby, the stability control re-uses low-level basic actuator access implemented both by the emergency driving function and the prototype driving function to control the vehicle.

The hardware units have to provide the basis for execution of the different functions outlined in the functional architecture. For the hardware architecture, the principle to keep components independent and the requirements for modularity of the vehicle are regarded. Figure 8.7 outlines the resulting architecture on system level. The vehicle state acquisition function and the stability control function are implemented on independent hardware units (*inertial measurement system* and *stability control system*) with no redundancy. A failure of one of these units can not induce a total system failure if fail silent behavior is ensured. At least, the emergency driving

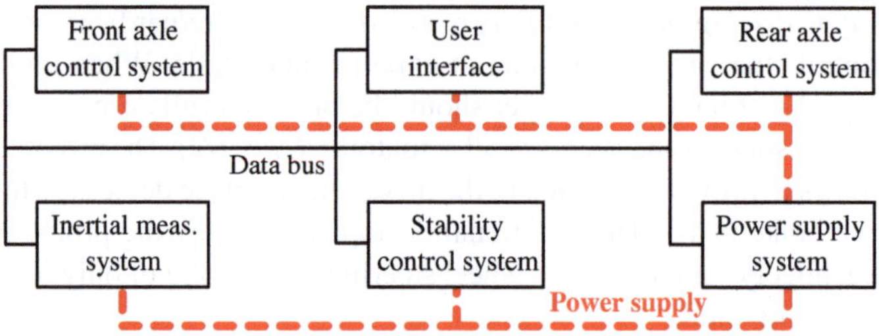

Fig. 8.7 Hardware architecture of MOBILE on "system layer"

function will be available. The driving functions (emergency and prototype) are distributed over three control systems. The *front axle control system* controls the actuators of the front axle and pre-processes all sensor data acquired at the axle. This contributes to fulfill the modularity requirements (*Requ. 1*). For modifications to the front axle, only the according control system has to be adapted. The rear axle control system is structured analogously. The *user interface* acquires the user inputs and provides the data to all other units via a *data bus* backbone. An additional simulation system can be added to the hardware set-up for more complex calculations required by prototype functions. As the simulation computer can always be disconnected from the network in case of failure, it is not further regarded in the hardware architecture. A *power supply system* provides the required electrical energy to all mentioned systems. The intelligent power supply system contributes to the desired fail-silent behavior of each individual system. More details will become obvious on lower hierarchical layers.

Figure 8.8 illustrates a merged view on hardware and functional aspects. It becomes obvious that the driving function and especially the diagnostic and safety related functions are distributed throughout the whole network. This supports the modularity concept as, e.g., axles can easily be exchanged as low-level tasks are kept local. Also, it reduces the probability of failure of the vehicle control function. In case of failure of a component, the system maintains at least a degraded mode of operation.

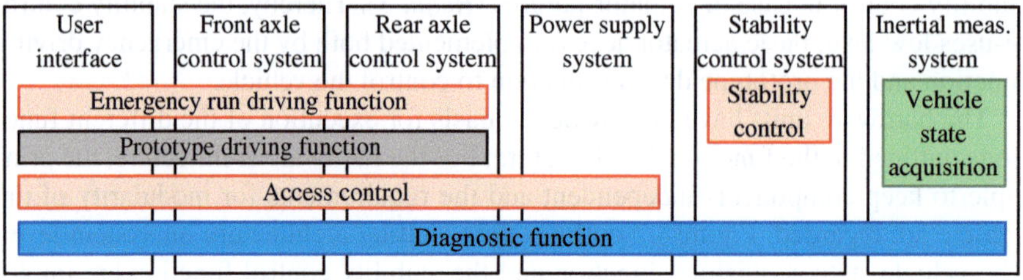

Fig. 8.8 Association of functional elements to hardware units on "system layer"

8.2.4.3 Subsystem Layer

Further detailing of the functional architecture on subsystem level reveals important functional modules and their interaction (Fig. 8.9). For the driving functions, *data acquisition*, *data processing*, and *actuator control* are distinguished. If necessary, the access control function blocks actuator access rather than starting or stopping the execution of a function. Thus, data acquisition and processing is performed in a hot-standby manner ensuring short switching times. The vehicle state acquisition is split into *inertial measurement* with an according sensor platform, *sensor data fusion* and *state estimation*. The sensor data fusion combines the inertial measurements with classical sensor data as wheel speeds or steering angles. Depending on the mode of operation, data from classical sensors is acquired from the emergency run function or the prototype driving function. The state estimation then takes the gathered and fusioned data as a starting point to estimate unmeasurable values as the side slip angle.

The stability control generates a *reference behavior* of the vehicle based on internal models of desired vehicle dynamics. If deviations in vehicle behavior from the reference given by the models are detected, the stability control modifies the actuator commands by the driver to ensure safe driving. Depending on the task of the stability control different reference models are referred to. For classical stability control, a model approximates the stable behavior of the real vehicle in order to identify critical deviations in vehicle behavior. Additionally, a simple front wheel steering vehicle model with limited dynamics resembles the fail-safe behavior of the vehicle in case of an actuator failure. Thus, the vehicle control system has to guarantee this minimal performance of the real vehicle even in case of a defined number of failures of actu-

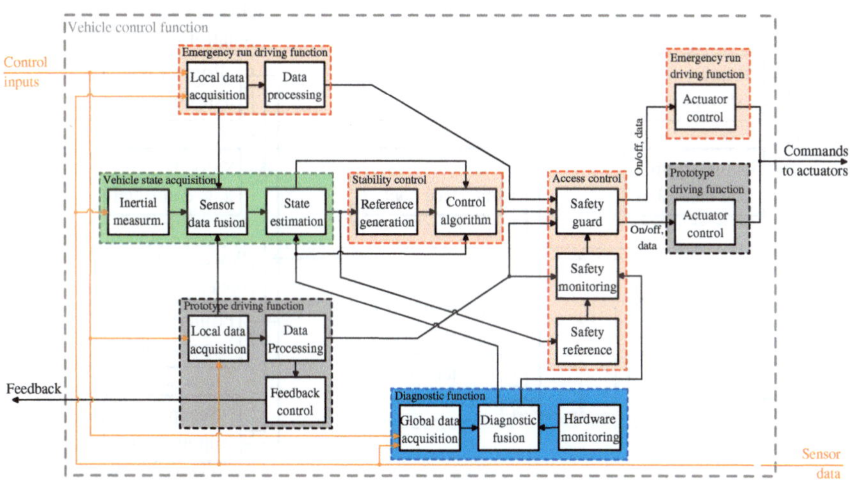

Fig. 8.9 Functional architecture of MOBILE on "subsystem level"

ators. This serves as a basis to describe the safe state of the vehicle in a later safety evaluation.

With the increased level of detail on subsystem layer, the information sources for the diagnostic system become obvious. The system extracts necessary information from *hardware monitoring* algorithms and the *globally available data* in the vehicle. Hardware monitoring relies on diagnostics as referenced in the state of the art (Sect. 8.2.3). The global data acquisition takes into account the driver commands and the reaction of the vehicle to these commands. If deviations between desired and actual vehicle handling become significant, the diagnostic system identifies possible failures. Still, the diagnostic unit has a more hardware and EE system focused perception of the vehicle compared to a stability control system. More details on the diagnostic algorithms can be found in Bergmiller et al. (2011b).

The access control is split into *safety reference* generation, *safety monitoring*, and *safety guard*. These units re-configure the system if a driving function fails. Basically, the operation is similar to the one of a stability control: A safety reference describes the desired state of the EE system. The safety monitoring detects failures in the system behavior by comparison to this reference and commands system reconfiguration if necessary.

The hardware architecture on subsystem layer is outlined in Fig. 8.10. If only one failure has to be tolerated, the stability control and the inertial measurement unit need not be fault-tolerant and thus are implemented on only one controller each (*inertial measurement controller* and *stability controller*). Still, each unit is powered by both power supply lanes. This is necessary, as a total loss of one power line leads

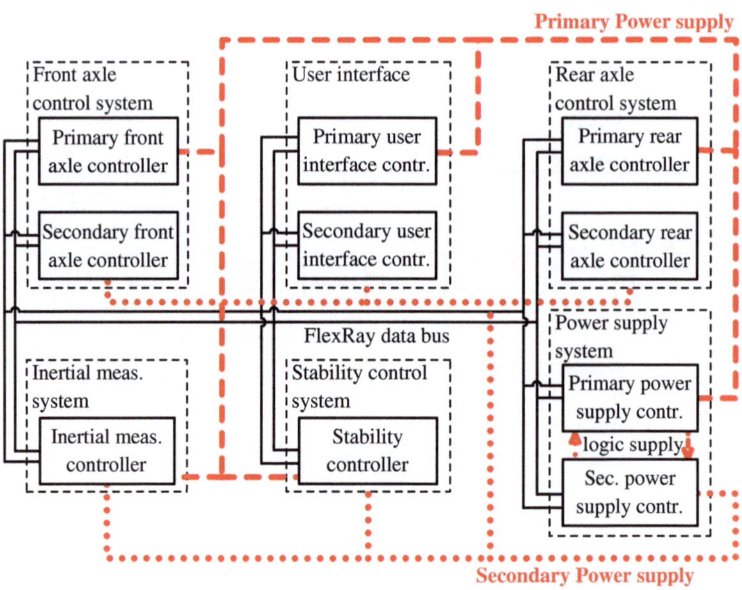

Fig. 8.10 Hardware architecture of MOBILE on "subsystem layer"

to a loss of all connected controllers, two diagonal brakes, propulsion at one axle, and two diagonal steering motors. Resulting, a stability control system is needed to maintain at least emergency operation of the vehicle. If the stability control and state acquisition would be powered by one lane, only the failure of one power lane would be manageable. The safety analysis of MOBILE showed that a total failure of one power supply lane is not possible due to a single point fault, but still a failure of the 12 V lane is possible and stability control is useful to compensate the resulting loss of two brakes and two drive motors. If the stability control itself fails, power can be cut by the power supply system and fail-silent behavior is achieved.

A failure of all other systems would lead to a full or partial loss of control of the vehicle. Thus, these systems are set up as fault tolerant units consisting of two redundant controllers each. The two controllers within a fault tolerant unit are powered by different power supplies that are controlled by individual *power supply controllers*. Resulting fail-silent behavior can be assured for each component. Communication between the modules is performed via a fault tolerant and time triggered FlexRay data bus with physically separately wired redundant channels. The data bus communication is designed to support in-cycle response and a cycle time of 4 ms. According to the information given in Sect. 8.2.3 and experiences in the project MOBILE, this timing suffices for stable operation of the vehicle and to implement high performance control algorithms for vehicle dynamics. Redundant information that is transmitted within one cycle is distributed equally over the communication cycle to reduce the impact of burst errors. Within each fault tolerant unit, sensors for basic actuator control or to acquire driver inputs are available in triple redundancy for majority voting, while actuators are only available once for each task. The wiring of the sensors and actuators to the axle controllers is done according to the requirements of the components available on the market. Mostly, CAN-bus connections implementing a CANopen protocol are relied on. Furthermore some digital and analog signals of sensors are evaluated and directly connected to the axle controllers. Within each axle, the allocation of sensor signals and actuator commands to bus systems ensures that the stability control unit can continue to control the axle such that at least neutral behavior with regard to vehicle dynamics can be achieved in case of a failure of a bus connection. E.g., the drive motor of one wheel is connected to a different bus than the braking unit of this wheel. Thus, in case of failure of one of the systems, the wheel can still be decelerated to some extent—either by recuperative braking or by mechanical braking. For MOBILE, research is ongoing to furthermore clarify potentials and limitations of control algorithms to handle actuator failures (Goldschmidt 2012; Lieberam 2011; Töpler 2010).

To avoid loss of control due to loss of power, the power supply system is set up redundantly. The power supply controllers contribute to the desired fail-silent behavior of all components in the vehicle. If a controller is classified as defective, the power supply for the controller can be cut. Thus, fail-silent behavior can be enforced externally if internal mechanisms fail. Within the power supply system, the two power supply controllers supply the logic part of each other. As a result, a malfunctioning power supply controller can be switched off by the partner controller. The safe state of the controllers is to supply all connected controllers in case of unpowered control logic. Using this configuration, all reasonable failure scenarios of the power supply can be handled in cooperation with decentralized safety guards executed on each node.

Again, the hardware and functional view are merged to determine the allocation of functions to hardware components (Fig. 8.11). It becomes obvious that the front and rear axle modules are set up analogously. The controllers within each according fault tolerant unit perform different tasks. The primary controller is intended to execute the prototype software under development and control all actuators including feedback generation. The primary controllers do not significantly contribute to monitoring of safe vehicle operation. Only the hardware monitoring algorithms provide feedback on the state of the node via the network. Additionally, a simple safety guard can block the boot of the node if an according command is received from the power control unit. In case of a necessary intervention of the stability control based on the primary controllers, base access to the actuators is granted. These algorithms are not visible to the user and are executed in the background. This allows the developer to act almost unrestricted by the safety concept.

The main diagnostic functions are implemented on the secondary controller. These controllers are operated in hot-standby manner to take over vehicle control if required. While not performing vehicle control, the available resources are used to perform sophisticated diagnostic algorithms. These algorithms continuously compare the behavior of the primary controller to a safe reference. Furthermore, all inputs from

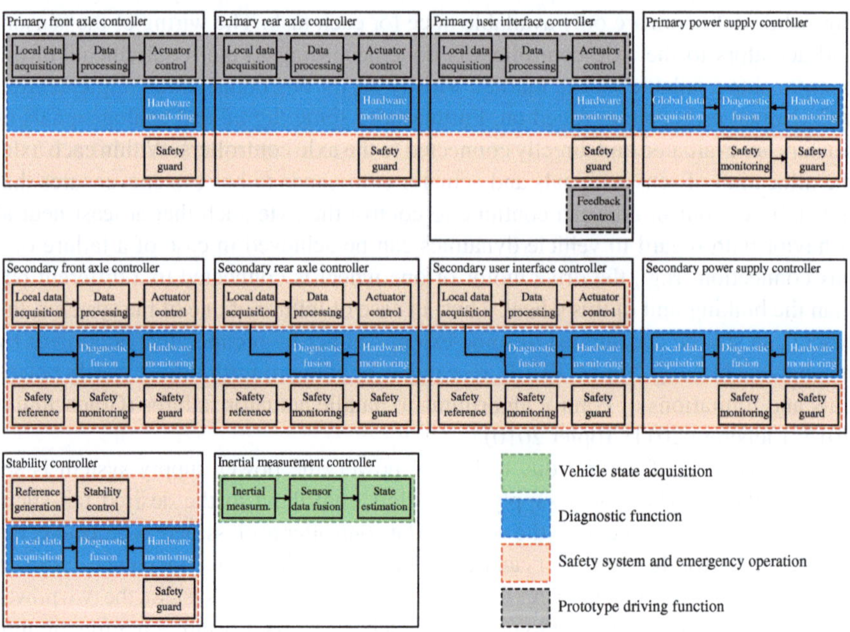

Fig. 8.11 Association of functional elements to hardware units on "subsystem layer"

the hardware monitoring of both primary and secondary controller are regarded. In case of failure of one node, the secondary controller communicates the failure state to the power supply controllers—either by a dedicated message or by falling silent. The power supply controllers can then command one node to transition into fail-silent mode and the other to maintain or take over the driving task. Also, a defective node can simply be disconnected from the power supply. It is important to note, that the secondary controller continually has to identify the basic behavior of the primary controller in order to achieve smooth taking over after a failure. Thus, the steering ratio is adapted online. Still, the adaptation is performed within strict bounds to avoid adaptation to erroneous behavior. If the secondary node is subject to failure, a restart or reinitialization can be triggered to repair the system. For the primary node, such measures are not applicable as the prototype software might go through unintended initialization routines while driving.

Obviously, the secondary controller represents a possible source of single point failures as it is mainly responsible for the diagnostic tasks and thus the decision making within one axle. This challenge is addressed by the self-monitoring of each node and monitoring of each node by the remaining network. The distributed diagnostic system on all nodes monitors proper operation of all other nodes by an Alive Network Management Vector based mechanism similar to the one implemented by TTP/C on bus controller level or (Sakurai et al. 2008) as an extension to OSEK. The alive monitoring is implemented on application layer and allows to derive the operability status of each node within the network on application level. Thus, advantages of a network centric architecture are exploited.

As already indicated, the power supply controllers play an important role for the central coordination of the safety concept. Basically, the access control is mainly implemented on the power supply controllers. To prevent scenarios where a power supply controller generates single point of failure scenarios, the power supply nodes monitor each other intensely and include information from the network wide alive monitoring. Additionally, the individual nodes in the axle modules perform consistency checks between the commands of the two power supply units. For the first version of the safety concept, the overall logic is based on the assumption that only one independent fault has to be tolerated. All multi-point faults are assumed to lead to a loss of vehicle control.

With the presented merged view on subsystem layer, the introduction of the hardware and functional architecture concludes. The following hierarchical layers (component and elementary layer) feature further increasing levels of detail and focus on individual functions and their allocation to hardware parts within one controller. To some extent, these investigations are made within the project MOBILE during layout of the electronic components and software implementation. Still, several hardware parts are sourced externally and no detailed information is available.

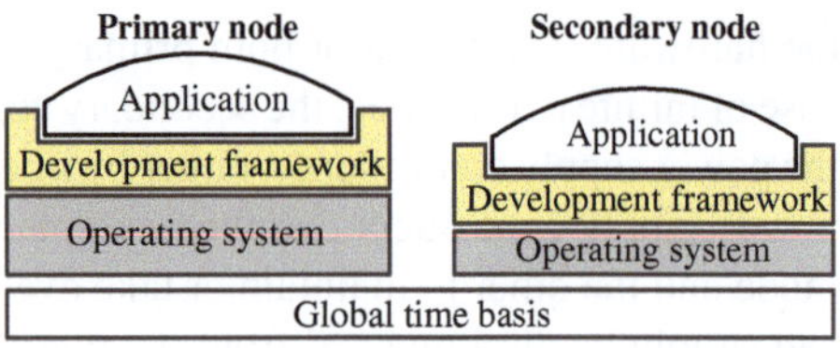

Fig. 8.12 Time synchronization and software framework on the network nodes

8.2.4.4 Software View

Complementing the architecture introduction, a brief look at a software view of the system is taken (Step 6 in Fig. 8.1). As mentioned, the software view of the system on the individual layers widely correlates with the introduced functional architecture. After all, each elementary function can be implemented as a software function provided by an object or a dedicated stand-alone function. Still, the overall software structure "orthogonally" to the demonstrated application layer was not yet regarded. This structure includes the layered approach from hardware abstraction to the application software modules and the organization of their interaction. For the project MOBILE, Fig. 8.12 provides a simplified overview of the structure on each node. Each node runs a custom written operating system that fulfills the requirements for task scheduling, minimal resource consumption, and latencies while featuring the required flexibility and allowing full access to all components. Also, it ensures that the node synchronizes itself to the global time basis based on the FlexRay bus clock. Resulting, all actions within the individual nodes can be synchronized at a precision of microseconds if necessary. This way, also in-cycle response and just-in-time data processing and transmission can be realized. The operating system interfaces with the development framework provided by Mathworks Embedded Coder. The application modules are then integrated as Simulink blocks that are written in C/C++ code or modeled graphically with Simulink. This strongly facilitates code reuse and reduces coding errors due to graphical programming. The application modules can—if required—be executed on any node within the network due to the common interfacing. To reduce common cause failures, primary and secondary node run different operating systems. Summarized, the presented layered approach allows flexible exchange and reuse of software modules while at the same time hard real-time requirements can be met on microcontroller hardware (*Requ. 2*).

8.2.5 Summary and Criticism of the Presented Architecture

In the previous sections, a by-wire architecture for an experimental vehicle was derived in a top-down manner as far as possible within a research project. The architecture bridges the gaps (a) between flexibility and safety by network based monitoring combined with degradation concepts and (b) between safety and costs for hardware redundancy by exploiting functional redundancies. Thus, the architecture can fulfill the initially set requirements on flexibility, safety, and reduction of hardware redundancies (step 7 in Fig. 8.1). Several key aspects distinguish it from other approaches:

- The vehicle control function is implemented as a highly integrated system of all driving functions keeping required hardware redundancies low (Sect. 8.2.3) while allowing to fully exploit cross couplings between different functions. Comparable systems with focus on safety and a similar actuator set up are not found in literature.
- The architecture strongly emphasizes the importance of distributed execution of safety critical diagnostics on application level to achieve low failure rates while keeping the degree of redundancy low.
- System degradation including exploitation of functional redundancies is an indispensable part of the architectural approach. Online reconfiguration and "online repair" of components, e.g., by reinitializing components is possible. Still, repair mechanisms are not yet regarded for safety analysis.
- All actuators installed in the vehicle are used to generate novel functionalities while also contributing to system safety. Redundant actuators are economized.
- A cost efficient mean to achieve fail-silent behavior of the individual network nodes is implemented based on joint action of the power supply controllers and decentralized safety guards.
- The architecture supports flexible development of prototype software on the primary controllers with little restrictions due to safety mechanisms. Sophisticated monitoring algorithms help to keep safety mechanisms out of the application software and perform external monitoring. This approach might also be extended to complex functionalities in series vehicles. Then, not the complex function itself, but the external monitoring system has to comply with the given safety requirements. If done properly, this external safety guard can be structured generically and be re-used for different versions of the complex function.
- A model of the intended vehicle handling clearly defines the emergency run of the vehicle. This model serves as a baseline for benchmarking stability control algorithms but also for functional safety analysis. Resulting, well-defined requirements for proof of functional safety are defined moving away from less meaningful requirements on the behavior of individual components as used so far.

The architecture is based on important assumptions that are partially still subject to research. It is assumed that the FlexRay system with redundant channels suffices to fulfill the automotive safety requirements. Still, detailed safety investigations for data bus systems are ongoing. If required, the data bus system could be exchanged or extended. Moreover, the stability control to exploit functional redundancies represents the key for the reduction in redundant actuators and thus the key to one of the main benefits of the presented architecture. Based on the results of multiple research projects in vehicle dynamics and investigations in the project MOBILE, it seems reasonable that the stability control will be able to handle failures of individual actuators. Still, so far hardly any quantitative investigation of cross-compensations between dif-

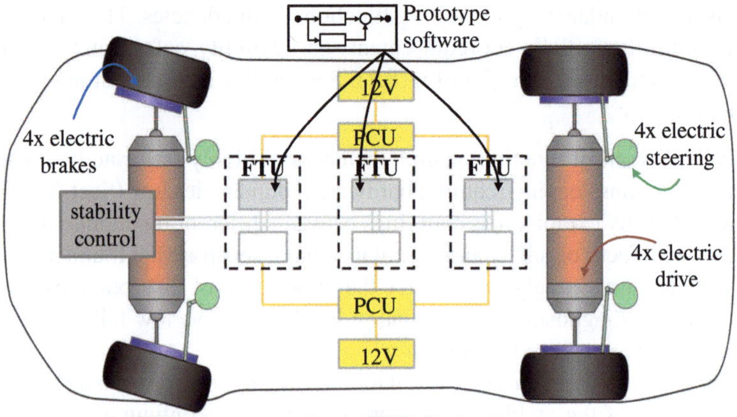

Fig. 8.13 Simplified architectural overview of MOBILE

ferent types of actuators under varying environmental conditions are available in the research community.

Summarized, the architecture enables construction of a powerful development platform. On the one side, novel applications for highly flexible vehicles can be evaluated during real test runs. On the other side, new means to ensure safety based on functional redundancy and vehicle stability control can be developed and verified. Concluding architecture derivation, Fig. 8.13 summarizes important aspects of the vehicle architecture merging mechanical, hardware and software views.

8.3 Functional Safety Evaluation

> A person who, intentionally or negligently, unlawfully injures the life, body, health, freedom, property or another right of another person is liable to make compensation to the other party for the damage arising from this.
>
> German Civil Code[17]

The above excerpt of the German Civil Code transfered to the automotive industry stresses the duty of any car manufacturer and engineer to ensure that its products are designed according to the state-of-the-art in terms of safety. If a vehicle demonstrably fails due to negligent design, the responsible person can be held liable for any consequences and thus has to provide according compensation. As mentioned in Sect. 8.1, the proof of a state-of-the-art design of modern vehicles on functional

[17] Official Translation by the Langenscheidt Translation Service of the German Civil Code (BGB) §823 in the version of its promulgation from 2nd of January 2002, last amended by statute of 28th of September 2009.

level is becoming more and more challenging for the car manufacturer. To give a guideline for proper design and safety validation of new vehicles on functional level, the ISO 26262 for functional safety in vehicles was derived from the more general IEC 61508[18] for functional safety in electronic safety-related systems. As a result, the ISO 26262 also serves as benchmark for the design of a vehicle and the design process in case of law suits. One key aspect of the ISO 26262 is the hazard analysis and the resulting classification of the derived safety goals in terms of ASILs.[19] Amongst others, these levels determine upper thresholds for the acceptable failure rate of the investigated function. Especially, in the field of electric vehicles or by-wire approaches, ISO 26262 opposes high demands on newly developed safety critical systems. These systems do not have a long lasting history with associated statistics based on millions of sold vehicles to rely on, but safety of the overall system has to be proven in full compliance with ISO 26262.

This section introduces an approach for hierarchical safety analysis of the "driving functionality" provided by a drive-by-wire vehicle with close functional couplings between individual units. Thereby, especially functional cross compensations between highly safety critical systems that are traditionally investigated separately (e.g. braking, steering, and drive system) are exploited for proof of functional safety. MOBILE serves as suitable demonstration platform as it features a high degree of functional redundancies (Sect. 8.2). In series vehicles, such redundancies are increasingly being introduced. Possible configurations include independently controllable rear axle steering (Pruckner et al. 2012) or superimposed steering systems at the front (Pfeffer and Harrer 2011; Pruckner et al. 2012) and full or hybrid electric drive train structures that allow torque vectoring. Still, cross-actuator functional redundancies are a topic of research and not yet investigated for series vehicles in the context of verification of functional safety.

8.3.1 Summary of Results of the Hazard Analysis

For the vehicle control function of the experimental vehicle MOBILE, a hazard analysis was performed based on ISO 26262. The procedure given in ISO 26262 is adapted to suit the evaluation of an experimental vehicle with high functional integration that is developed from scratch. Thereby, it is argued that (a) there is a strong relation between the evolving system architecture and the hazard analysis that requires iterative re-evaluation of hazards. Additionally, the approach proposed by ISO 26262 is (b) adapted to suite the special conditions of operation of the experimental vehicle. E.g., special circumstances due to operation on a closed off test track with well known environment are taken into account during safety analysis.

[18] IEC 61508: Functional Safety of Electrical/Electronic/Programmable Electronic Safety-related Systems, edition 2.0.

[19] Automotive Safety Integrity Levels (ISO26262-1:2011, p. 2).

Table 8.3 Assumptions for operation of the experimental vehicle on the test track

Assumption 1	A **skilled test driver** is driving the vehicle. The driver is capable of handling critical driving situations on a high friction surface if sufficient means to control the vehicle are provided.
Assumption 2	Test runs are only executed in **good weather** (dry road, no rain).
Assumption 3	The driver wears a **protective suit** as, e.g., used in formula one vehicles.
Assumption 4	High speed tests are only performed on a **wide open testing ground** that allows the vehicle to come to a safe halt even if the braking system fails. Thereby, it is assumed that the drive motors can be deactivated by the driver.
Assumption 5	Critical sections of the test track denote sections where the closest **buildings or obstacles** are located at a **minimal distance** of 6 m orthogonally to the track.
Assumption 6	The experimental vehicle features an **emergency-off system** that allows the driver to cut the power of the drive motors at any time.
Assumption 7	For critical sections of the test track, a **speed limit** of 10 m/s is set that has to be obeyed by the test driver. Combined with Assumptions 5 and 6, a worst case impact speed into obstacles in case of a failure of approx. 13.9 m/s (50 km/h) results.[a]

[a]This speed was determined based on simulation experiments with a double track vehicle model, as, e.g., described by von Vietinghoff (2008), assuming different steering concepts, distances to obstacles ranging from 6 m to 10 m orthogonally to the track, a high friction surface, unintended acceleration of the drive motors, and a reaction time of the driver to hit the emergency off of 0.6 s. This reaction time corresponds to typical reaction times of well-trained average drivers, e.g., for emergency braking (Mclaughlin 2007; Mehmood and Easa 2009). The given maximal impact speed was determined for a scenario where the steering angle at the front and rear axle were set to 0.35 rad and 0.09 rad in equal directions.

To start with, Table 8.3 provides an excerpt of important assumptions that were made for the operation on the test track. The given hazard analysis is only valid as long as the vehicle is operated under these conditions. The safe state[20] of the vehicle in case of any failure is defined as follows:

> The vehicle continues to provide basic means for vehicle control to the driver until the vehicle can be safely halted.

Thereby, a linear vehicle dynamics model with front wheel steering and drive that is evaluated online defines the required minimum performance of MOBILE in case of a failure. This emergency operation has to be maintained for a given time interval. More details will be given in Sect. 8.4.

Now, Hazards while driving MOBILE on critical sections of the test track are identified. Therefore, the approach given in ISO 26262 for series vehicles is followed. The results are given as hazards that are evaluated in terms of ASILs[21] depending

[20] ISO 26262 defines the safe state as "the operating mode of an item without an unreasonable level or risk" (ISO26262-1:2011, p.14), while risk refers to the "combination of the probability of occurrence of harm and the severity of that harm" (ISO26262-1:2011, p.13).

[21] The Automotive Safety and Integrity Levels (ASILs) are used to classify hazards according to ISO 26262. ASIL levels range from A (least stringent) to D (most stringent).

Table 8.4 Excerpt of the modified hazard and risk assessment according to ISO 26262

Hazard	Severity	S[a]	Probability of exposure	E[b]	Controllability	C[c]	ASIL
Unintended acceleration leading to a crash	Survival of the driver is uncertain as collisions at high speed are possible.	**S3**	Frequent operation of the vehicle at locations where unintended acceleration can cause collisions	**E4**	A skilled test driver can simply control the vehicle by applying the emergency-off system and/or brakes.	**C1**	**B**
Deviation from the yaw rate reference intended by the driver leading to a crash	Light and moderate injuries are likely at low speeds ($\leq$10 m/s) for a driver wearing a protective suite.	**S1**	Frequent operation of the vehicle at locations where deviation from the yaw rate reference can cause collisions	**E4**	At the given low speed, a skilled test driver can normally control the vehicle by braking.	**C3**	**B**

[a] The levels S0 to S3 classify the severity of an accident. S0 denotes lowest and S3 highest severity
[b] The levels E0 to E4 classify the exposure. E0 denotes lowest and E4 highest exposure
[c]The levels C0 to C3 classify the controllability. C0 denotes best and C3 worst controllability

on the expected controllability,[22] severity,[23] and exposure.[24] Table 8.4 outlines two exemplary hazards that will be used as a reference in the following.

These hazards are then associated to safety goals. Safety goals again serve to derive technical and functional safety requirements for system components. Each safety requirement inherits the ASIL classification from the safety goal and thus from the identified hazards unless ASIL decomposition[25] is applied. If this is done for a highly integrated drive-by-wire system as introduced for MOBILE (see Sect. 8.2), one notes that all requirements on vehicle level that are not associated to the emergency-off system have to be associated to the overall vehicle control function. An association of safety requirements to clearly separated sub-functions may be possible in the functional architecture but is no longer useful if the hardware architecture is taken into account. E.g., one hardware unit contributes to braking, steering, and propulsion function. Figure 8.14 illustrates such a system structure on vehicle level for an

[22] Controllability refers to the "ability to avoid a specified harm or damage through the timely intervention of the persons involved, possibly with support from external measures" (ISO26262-1:2011, p. 4).

[23] In this context, the severity gives an "estimate of the extent of harm to one or more individuals that can occur in a potentially hazardous situation" (ISO26262-1:2011, p. 16).

[24] Exposure classifies the frequency of being in a "an operational situation that can be hazardous if coincident with [the currently investigated] failure" (ISO26262-1:2011, p. 6).

[25] According to ISO 26262 ASIL decomposition denotes the "apportioning of safety requirements redundantly to sufficiently independent elements" (ISO26262-1::2011, p.2).

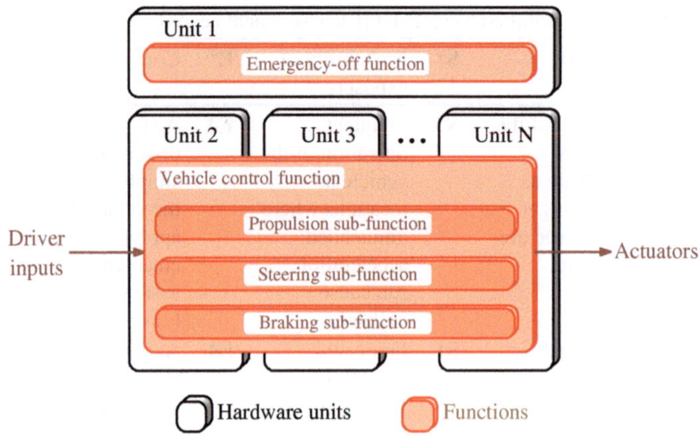

Fig. 8.14 Highly integrated demonstration system

experimental vehicle like MOBILE with emergency-off system and a highly integrated vehicle control system based on drive-by-wire. Thereby, a functional view with hardware units in the background is chosen. The given system consists of several individual hardware units that are combined to fulfill the overall task. A similar tendency towards integration of multiple safety critical functions on one control unit and within one mechatronic components can be seen in modern hybrid electric and electric vehicles: BMW proposes an "integrated chassis management" that closely links different control functions for longitudinal, lateral and vertical dynamics. Additionally, functions provided by one system are re-used by other systems, e.g., a head unit provides data that is used by several other systems as the active steering controller or the Dynamic Stability Control (Koehn et al. 2006; Smakman et al. 2008). Freitag and Kuhn (2012) go even further and replace conventional brakes at the rear axle with an in-wheel motor that drives and brakes the wheels exclusively. Thus, borders between different functionalities and classically separated systems start to blur also in series vehicles. A safety evaluation yields that the failure of one unit may lead to loss or degradation of multiple functions.

Transfered to the simplified system structure given in Fig. 8.14, the sub-functions merge into the overall vehicle control function. Thus, all safety requirements would have to be assigned to the one vehicle control function and the overall underlying hardware consisting of several hardware units. According to ISO 26262, the system would then be associated the highest safety requirements opposed on one of the executed functions. Resulting, safety evaluation according to ISO 26262 presented so far can lead to lower safety requirements on the overall system than intended: If one unit executes multiple functions of lower safety criticality that directly or indirectly effect vehicle handling, the overall criticality of this unit has to be higher than the one of each individual function. E.g., if the vehicle control function in a drive-by-wire vehicle fails, the driver has no means to intervene into vehicle control anymore. This renders the previous hazard analysis wrong and requires adaptation.

Thus, this contribution proposes that all hazards have to be re-evaluated iteratively based on knowledge about the evolving system architecture. Table 8.4 provides the results of this re-evaluation of the two example hazards. Thereby, only hazards are effected that require intervention by the driver by controlling sub-functions of the vehicle control function such as steering or braking. With the re-evaluation finished, the highest ASIL of one of the functional components of a unit can be assigned to the overall unit. For MOBILE, this yields a ASIL B classification of the vehicle control function that is taken as a reference for the safety evaluation introduced in the following sections. In this case, the re-evaluation did not increase safety requirements due to the independent emergency-off system and the well-defined environmental conditions. Still, the re-evaluation is necessary if the system architecture undergoes significant changes during development. Then, changes in hazard classification can occur that are not foreseeable at the beginning of the development project.

Note: It is important to note that the comparatively low safety classification of the experimental vehicle is based on the assumptions of a skilled test driver wearing a protective suit, the well known environment, and the emergency-off system. The emergency-off system features a serial redundancy structure of two emergency-off switches. The system is kept as simple as possible. Thus, it is very likely to be available to the driver in case of failures.

Remark for series vehicles: The hazard analysis for the steering or braking sub-functions would obviously yield an ASIL D classification for series vehicles. E.g., Richter and Köhnen (2012) and Sinha (2011) perform an analysis of these functions for electric vehicles with by-wire design confirming this result. Thus, the correct ASIL D classification of the overall system would already result without re-evaluation. This is intuitively comprehensible as ASIL D already represents the highest possible safety classification. Still, the need for re-evaluation pointed out in this contribution can be transfered to other highly integrated sub-systems in vehicles that execute several functions with lower ASIL levels.

8.3.2 Hierarchical Approach to Safety Analysis

The previous section demonstrated requirements on functional safety for highly integrated vehicle systems. Although, integrated drive-by-wire systems with multiple actuators and without mechanical or hydraulic fall-back layer as referenced in the exemplary hazard analysis are not yet available in series vehicles, the current tendencies in evolution of EE systems indicate similar challenges. High integration of functionalities promises enhanced customer benefit while limiting production costs, and by-wire control can serve as an enabling technology for further progress (Pruckner et al. 2012). The safety evaluation of such systems becomes time consuming and prone to errors (Papadopoulos et al. 2001). To address this challenge, the following

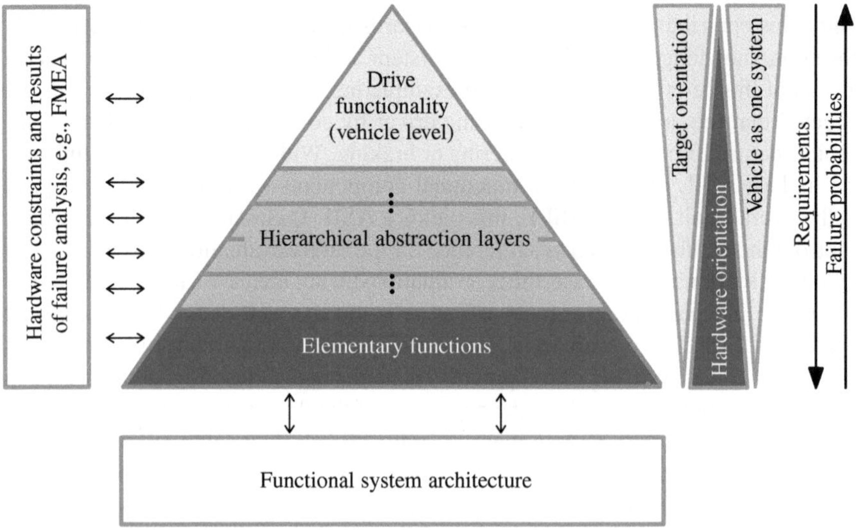

Fig. 8.15 Hierarchical approach to safety analysis

section introduces a tailored method[26] for hierarchical safety analysis that extends the existing approaches by several points: The hierarchical approach especially focuses on highly integrated systems with a *high degree of functional redundancies* and exploits these redundancies for safety purposes. These redundancies are currently hardly addressed at all for quantitative safety assessment. To reduce work effort for the analysis, the presented method introduces virtual systems and generalized failure states that allow to *focus the analysis on necessary components* by front loading knowledge on dependencies in the system. The hierarchical approach results in a *failure rate for the overall system* including the associated *emergency operation interval*. Knowledge about the available emergency operation interval is vital to ensure a safe stop of the vehicle and can possibly be exploited to define a limp-home mode. To assess the performance of the distributed diagnostic algorithms in the vehicle, the approach furthermore provides the diagnostic coverage of a globally operating *virtual diagnostic unit*. This diagnostic coverage can guide further development of the local monitoring algorithms or can be used for safety evaluation while at the same time encouraging "vehicle level thinking". To demonstrate applicability of the method, a prototype tool environment was set up, and the design of MOBILE was evaluated.

Figure 8.15 illustrates the perception of the investigated system adopted by the hierarchical approach. The safety analysis is performed on different hierarchical levels for the overall vehicle and its components. Quantitative results and failure probabilities are propagated bottom-up, while dependencies of components and

[26] A method is "a way of proceeding or doing something, esp a systematic or regular one" Collins (2010). During a development process, (multiple) methods can be applied to achieve necessary results (Hammerschall 2008).

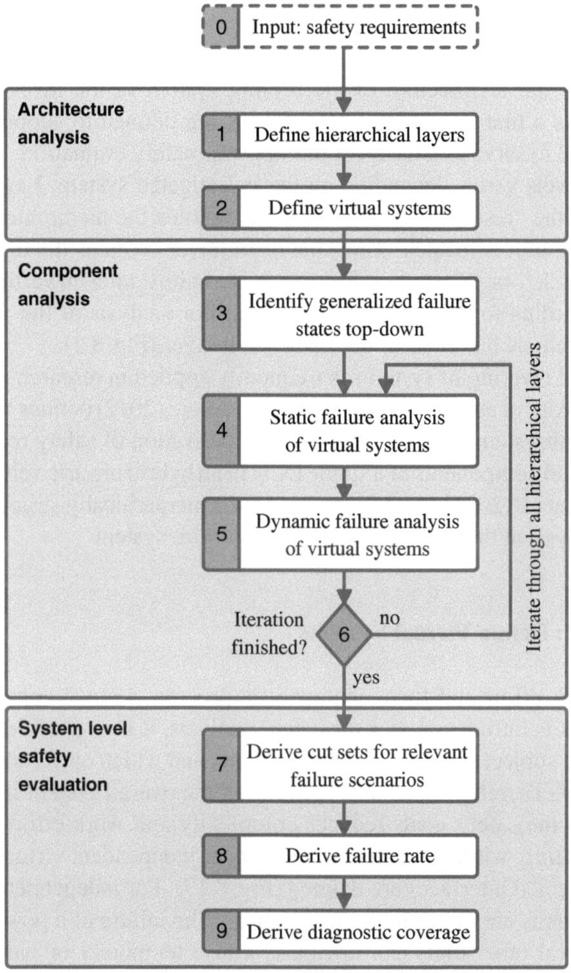

Fig. 8.16 Important steps of the hierarchical approach to safety analysis

requirements are forwarded top-down. The functional structure of the investigated vehicle and the constraints due to the hardware and software architecture are regarded on all hierarchical layers. Thereby, the hardware influence diminishes with increasing hierarchical level, but therefore the required understanding of the overall vehicle by the person in charge strongly increases. Both, profound knowledge about interconnection of vehicle components and knowledge about vehicle dynamics are highly relevant on higher hierarchical levels.

Figure 8.16 outlines the required steps for the hierarchical approach. In general, the process starts with a targeted analysis of the vehicle architecture then investigates relevant components and finally performs the evaluation of functional safety of the overall system. The following sections detail the individual steps and provide according related work.

8.3.2.1 Step 1: Define Hierarchical Layers

To begin with, the architecture of the driving system of the investigated vehicle is analyzed. As a first step, hierarchical layers are defined to support handling of complexity and to serve as a basis for the stepwise safety evaluation. The number of investigated levels varies depending on the investigated system. Layers have to be detailed up to the "result layer". Result layer denotes the hierarchical layer where the investigated unit is located. Thus, the result layer contains the unit that shall be classified as "o.k." or "not o.k.". For this unit, failure rates have to be given and evaluated according to the ASIL classification. For analysis of the vehicle control function, the vehicle layer has to be set as result layer (Fig. 8.2).

Hierarchical layering of systems is frequently applied in research and industry to handle complexity of automotive systems. E.g., Abele (2012) defines "vehicle level", "system and subsystem level" for hierarchical derivation of safety requirements for subfunctions and components of a single ECU in an hybrid electric vehicle. Similarly, Papadopoulos et al. (2001) identify the need for a hierarchically structured approach for safety analysis at the example of a brake-by-wire system.

8.3.2.2 Step 2: Define Virtual Systems

To reduce work effort and focus the analysis process, a novel approach based on virtual systems is introduced. For the safety analysis, it is vital to determine which subsystems are subject to common cause failures, and which ones can be assumed to be independent. Thereby, reasonable splitting of the overall system into subsystems with regard to the safety goals reduces complexity and work effort for the safety analysis. Resulting, within each hierarchical layer independent virtual systems with clearly documented interfaces are defined (Fig. 8.17). For independence of systems, power supply units are of particular importance. The failure of a power supply connected to several other units can obviously cause temporary or partial loss of all supplied units due to over- or undervoltage. This needs not be regarded as a form of dependence at this point, but will be handled later on. If a dependence is assumed, the granularity of the safety analysis is reduced and the safety evaluation becomes more pessimistic unless more effort will be taken in step 3 (explanation given there).

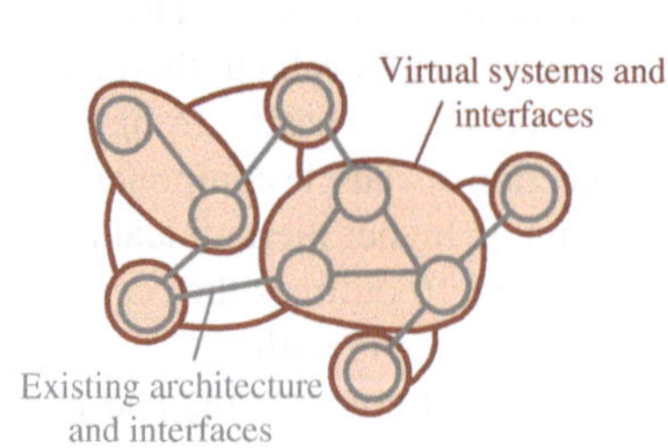

Fig. 8.17 Virtual systems introduced on one hierarchical level

The same strategy can be followed if other units for some reason have significant but similar effect on several other units.

The definition of virtual systems is challenging as the developer requires profound knowledge of the functional, software, and hardware architecture at the given hierarchical level. Still, all following steps of the analysis can then be performed based on the architecture of virtual systems without having to regard other architectural perspectives. The virtual systems ensure linkage between relevant views on the architecture, and overall complexity is reduced by front loading this knowledge. Typically, the definition of virtual systems will be performed middle-out. On the level of control units, independence between different units can be determined easier. Starting from there, the investigations are continued up- and downwards. Thereby, the boundaries of the virtual systems have to remain consistent throughout all hierarchical layers. A system on a lower layer must only be contained in one system at the next higher hierarchical layer.

If on a higher hierarchical layer (unless result layer) a split into independent virtual systems is not possible, this indicates that too many hierarchical layers were introduced or a weakness of the chosen system architecture was identified. The consequences will become obvious in step 7 and will be discussed in more detail there.

8.3.2.3 Step 3: Identify Generalized Failure States Top-Down

Starting with Step 3, failure modes of the virtual systems identified during the architecture analysis are investigated. As a basis for the analysis, the hierarchical approach introduces generalized failure states for the virtual systems defined on each hierarchical layer. These states abstract information on the current failure state of the virtual system. Thereby, only information needed by other systems on the same and especially on higher hierarchical layers is included. This significantly reduces work effort for later quantitative system safety evaluation when compared to existing approaches as, e.g., the HiP-HOPS approach proposed by Papadopoulos et al. (2001). Figure 8.18 provides an example for generalized failure states of a simple system. For the hierarchical approach, these generalized states serve as a well-defined and well-documented interface between experts or suppliers working on different subsystems throughout hierarchical layers. After definition of the generalized failure states, an expert working on a component can focus on a locally well-defined work package, while vehicle level effects are implicitly taken care of.

Still, the definition of the generalized failure states is a challenging task and requires cooperation among experts. It mainly follows two strategies: On the one side, the requirements from a higher layer have to be propagated top-down. This ensures that each state provides sufficient information to an expert working on the higher layer. The expert can then evaluate the overall system based on the pool of generalized failure states from the next lower layer. On the other side, the structure of the investigated virtual system influences the definition of the generalized failure states. Therefore, a rough understanding of the systems purpose and behavior in case

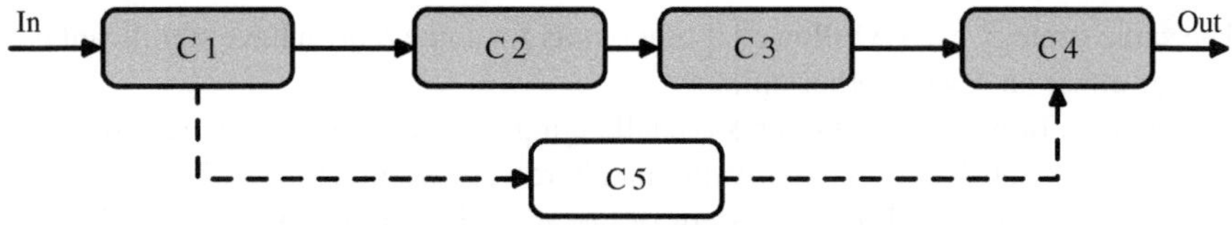

Failure of	Generalized failure state
C1 | C4 | (C5 & (C2 | C3))	Total loss of system
C2 | C3	Emergency operation
No failure | C5	Regular operation

Full operation: ⟶, Emergency operation: - ⟶, Logic "or": |, Logic "and": &, Component x: Cx

Fig. 8.18 Simplified example of a definition of generalized failure states

of failure is required. As a result, the generalized failure states of a virtual system have to be defined iteratively in cooperation between the affected experts. Typically, the necessary states should be pre-defined top-down to ensure target orientation and then be detailed by an expert for the investigated layer.

After definition of the failure states, each state is assigned a severity top-down-judging from the effect of the failure on the overall system. The original severity on vehicle level stems from the hazard and risk analysis performed according to ISO 26262. The assignment of severity levels is a vital input for the failure analysis of the subsystem detailed in the next step.

To reduce complexity, generalized failure states have to be targeted at safety goals. Unnecessary states, especially in lower layers, increase work effort for the analysis process. Also, too many states for a component can indicate insufficient granularity during definition of virtual systems. "Global failures" that effect several units in a similar way, as a loss of power, should be treated within a generalized failure state of the responsible unit. This procedure is well suited to, e.g., describe the effects of a loss of a central power supply unit.

In literature, the method of introducing generalized failure states is regarded to some extent so far: Sinha (2011) defines generalized failure states for a braking system regarding one hierarchical layer. The states are not exploited for linking systems or to hierarchically propagate severity levels. An application of generalized failure states to a more complex system is outlined by Rehage et al. (2005). The introduced states are identical for all systems ("active", "isolated", "active-hot", "passive-warm", "passive-cold") and applicable for aerospace systems with multiple parallel redundancy but are hardly compliant with the requirements to exploit functional redundancies in this automotive project.

8.3.2.4 Step 4: Static Failure Analysis of Virtual Systems

With regard to the generalized failure states, a more detailed analysis of the virtual systems is required. To start with, step 4 performs a "static" failure analysis without taking any timely effects into account. This represents the first step of a bottom-up iteration via steps 4, 5, and 6 (Fig. 8.16).

The failure analysis in step 4 targets at internal failures of virtual systems. On higher levels, the analysis is supported by the results from lower layers, as the investigated system consists of a defined number of already analyzed systems with associated generalized failure states. Different methods support the failure analysis: ISO 26262-4 suggests deductive, e.g., Fault Tree Analysis (FTA) as well as inductive analysis approaches, e.g., Failure Mode and Effect Analysis (FMEA). Details on FTA, FMEA, and further methods are, e.g, given by Rausand and Hoyland (2009) or Löw et al. (2010). This work mainly relies on a slightly modified FMEA that includes possible ways to diagnose and handle failures and a simplified FMEDA (Failure Modes, Effects and Diagnostic Coverage Analysis) to determine quantitative data. For FMEDA, failure rates of software components (control and monitoring algorithms) are included. This extends the classical approach for failure analysis given in ISO 26262 that exclusively refers to hardware components for failure rates. To some extent algorithms are regarded by the demanded diagnostic coverage achieved by certain failure detection mechanisms. In general, software is considered to comply to ASIL requirements if the software was developed according to the guidelines given in the standard. Software is not investigated quantitatively. This approach may be valid for series vehicles with profoundly developed software components and little dependence on external influences. Still, the failure rate given for the vehicle according to ISO 26262 assumes perfect software and is after all only valid for the hardware set-up. For the experimental vehicle, the failure rate of software components has to be considered for two main reasons: Firstly, parts of the software running on the vehicle are prototypical and feature failure rates that are several orders of magnitude higher than the ones of hardware components. These failure rates have to be estimated roughly based on experiences made in previous research projects. Secondly, algorithms as for vehicle stability control are a vital part of the safety concept. These algorithms can not be expected to operate properly under all environmental conditions or for all input configurations. The system, by its design, may just not be able to handle some rarely occurring situations. A failure rate has to be assigned that most likely has to be derived from statistical data acquired with a similar system under similar conditions of operation of the vehicle.

The failure analysis performed during the hierarchical approach profits from the option to apply methods as the FMEA locally: Usually, FMEA has to include failures that globally effect vehicle control, making the evaluation challenging for an expert for the local component. The hierarchical approach allows to evaluate the effects of a failure with regard to the generalized failure states that were defined for the component. The severity needed for the FMEA is then based on the top-down propagated severity associated to the generalized failure state in step 3. As a result, the global context is taken care of. Again, the allocation of tasks to local experts is supported.

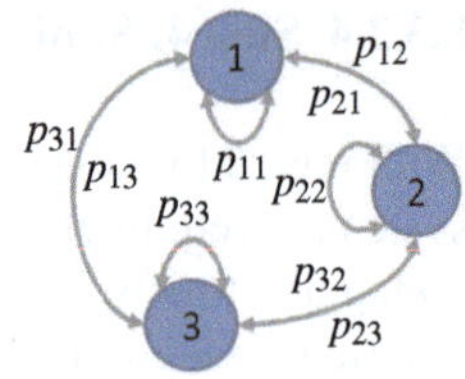

Fig. 8.19 Markov-Chain with three states

Additionally, the linkage of the severity analyses for different components via the top-down propagation supports comparability between results.

8.3.2.5 Step 5: Dynamic Failure Analysis of Virtual Systems

Following, for each available failure state of a system the probability of the system being in that state has to be derived. Therefore, an approach based on first order Markov-Chains is taken. This procedure has already been suggested by Tkachev for the general "analysis of systems with complex structure" in 1983 (Tkachev 1983) and was also followed by Zuo et al. (2005) to perform "quantitative reliability analysis [...] of steer-by-wire system[s]" in the automotive domain. ISO 26262-4 references Markov modeling in general as a valid way to analyze system design, too. A simple first order Markov Chain with three states is given in Fig. 8.19. The p_{ij} resemble the transition probabilities from state i into state j. According to Köhler and Broy (1983) the p_{ij} can be defined as:

$$p_{ij} = P(X_{t+1} = s_j | X_t = s_i). \tag{8.1}$$

Thereby, X_t defines the system state at the time step t and s_i resembles the feature vector characterizing the system state X within state i. As can be seen, the transition probabilities from one state into the other state at a given point in time only depend on the system state at the previous time step (Markov Property for first order Markov Chains). This is an important aspect for failure analysis as the history leading to a certain system state does not have to be known. All background knowledge has to be modeled by the structure of the Markov Chain. For failure analysis, the transition probabilities p_{ij} resemble failure rates λ of parts of a virtual system. These failure rates are derived from elementary hardware or software components on lower layers and are, by means of the hierarchical approach, propagated to any higher hierarchical layer similar to the approach presented by Papadopoulos et al. (2001). For the presented method up to two independent faults occurring one after the other are regarded for definition of the Markov Chain. This ensures that both the reaction of the system to the first fault and the mode of operation afterwards can be evaluated in step 6. Further consecutive faults are not regarded (see also recommendation of ISO 26262-5, Annex C), which reduces work effort. Failure rates and failure states of externally supplied components are directly fed into the system at the appropriate hierarchical layer.

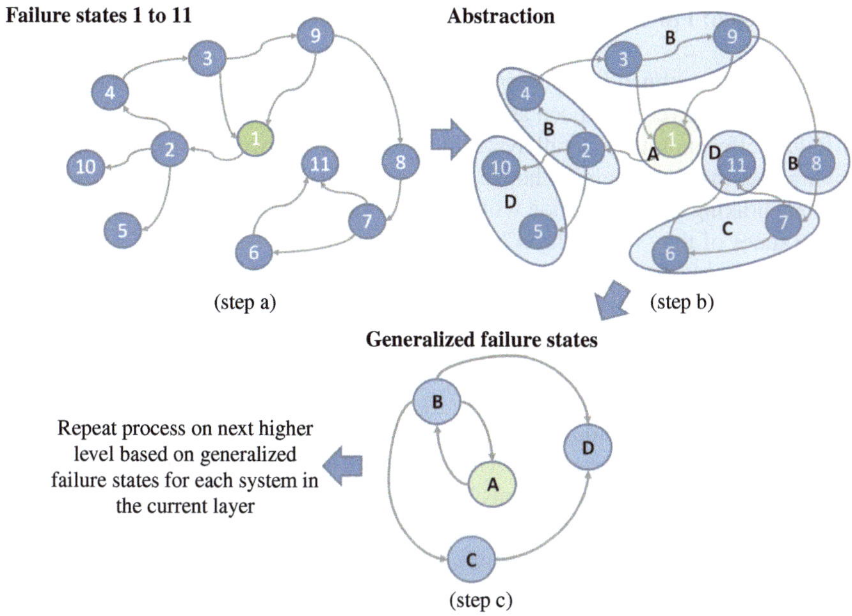

Fig. 8.20 Markov Chain and generalized failure states A to D for a system with failure states 1–11

As all failure rates that are associated to hardware components vary over lifetime due to aging, the Markov chain is not homogeneous and therefore is solved iteratively. The change of failure rates over lifetime of the vehicle is modeled separately. Ideally, the aging models rely on statistical data from systems already in the market. Otherwise, typical aging curves for components can be taken from literature.

After proceeding as outlined, a Markov chain results for each system at a hierarchical layer (Fig. 8.20 step a) with associated failure rates for state transitions. Figure 8.20 depicts only one way transitions as the system is assumed to be *not self-healing* for most failures. Thus, a re-transition from one failure state into a state with less failures is not possible unless, e.g., by a system restart leading through the "ok" state "1". The states of the detailed Markov chain will typically not yet resemble the generalized failure states of the system. The abstraction process is shown in steps b and c of Fig. 8.20. Thereby, states of the initially detailed Markov Chain are associated to the generalized failure states defined in previous steps. The failure rates for transitions between the generalized failure states are calculated from the underlying Markov chain by means of conditional probabilities and by summing up relevant transition probabilities.

8.3.2.6 Step 6: Finish Iteration through hierarchical layers

The evaluation outlined in the steps 4 and 5 is repeated until all hierarchical layers up to the result layer have been analyzed. In the result layer, the final classification of failure effects on the overall system has to be performed. For highly integrated driving systems, the developer has to asses the "vehicle control function" including steering, braking, and propulsion with regard to the effects of failures. The abstracted failure states of the "system layer" serve as input for the evaluation. Thereby, only single and double point faults are regarded:

At first, the developer one by one rates the effect on vehicle handling if a virtual system transitions into an abstracted failure state. The safe state for the vehicle as, e.g., defined for MOBILE in Sect. 8.3.1, serves as reference for minimal acceptable handling characteristics of the vehicle. In a second step, the developer assess the effects of an additional failure within the already faulty system or any other second virtual system. One by one, each virtual system is set to be in one of its failure states. Then, the possible transitions into generalized failure states of this and any of the remaining systems are investigated. Further state transitions do not have to be investigated. Assuming small failure rates, the probability of occurrence of even two independent faults within a given short time interval leading to various system failures is several orders of magnitude smaller than the probability of a single fault. In literature, it is demanded that only one independent fault has to be tolerated by the vehicle control system (Armbruster et al. 2006; Johannessen et al. 2002; X-by-Wire Project 1998). This is also backed up by the current legislative demands, e.g., for approval of the braking system for public traffic (ECE R13). For the safety analysis, it is assumed that more than two independent faults will always lead to a loss of vehicle control, which makes the analysis results slightly more pessimistic. The evaluation of the second fault is needed to determine, whether a sufficient emergency operation interval can be guaranteed to get the vehicle to a safe halt after a first failure occurred. Step 8 details the according calculations.

The above evaluations require profound knowledge of vehicle dynamics. Intense investigation of both the capabilities of actuators relevant for vehicle handling and the associated control algorithms is vital. Several research groups worldwide are investigating these effects and intense research focuses on the novel opportunities in electric or drive-by-wire vehicles. Some examples were given in Sect. 8.2.3.

8.3.2.7 Step 7: Derive Cut Sets for Relevant Failure Scenarios

Starting with step 7, the tool environment, set up for the hierarchical approach, automatically evaluates the so far gathered information numerically. The following two sub-steps are performed in step 7:

1. *Split system into an operational and a faulty part*: Each of the critical failure scenarios based on single or double point faults that were identified for "result layer" implicitly splits the system into an operational part and a faulty part that

is non-operational. The latter part causes the system to fail. Thus, the probability of failure has to be determined for this part. Prior to this, the system split has to be performed throughout all hierarchical layers associating each virtual system to one or the other part. Due to the hierarchical analysis performed so far, the linkage between the virtual systems is known within and between layers by means of the generalized failure states. Thus, the faulty system part can be defined automatically.

2. *Identify cut sets*[27] *for the relevant system part:* The algorithm can now identify all relevant faults and fault combinations that lead to a failure of the investigated part of the system. As already mentioned, only combinations of up to two faults are regarded in this analysis. Resembling the notion in reliability engineering, the algorithm identifies cut sets that lead to the relevant failure state. Thereby, each cut set is determined on a hierarchical layer that provides a complete set of quantitative data to evaluate the effected virtual systems. Typically, this will be the lowest hierarchical layer. Still, this approach also allows to easily integrate third party supplier components that are not detailed to lower hierarchical layers but provide failure rates at higher layers. If needed, the algorithm precisely indicates missing failure rates that have to be provided. Thus, only failure rates that are directly relevant for safety analysis are required. As generation of this quantitative data is costly especially for novel components, the front loading performed by the hierarchical approach due to virtual systems can reduce costs when compared to other approaches that start from a full set of quantitative data (Papadopoulos et al. 2001).

   Each identified cut set is then referenced according to the notion introduced in Fig. 8.21. The superscript for the layer l and the cut set counter i uniquely identify a cut set. The indices s and d are supplementary to highlight whether the cut set is based on a single or double point fault. They constitute redundant information for better readability and later reference. If one of the additional indices or superindices is not given, this references a number of cut sets with all valid combinations for the omitted indices. E.g., C^l resembles all cut sets of the hierarchical layer l and C alone resembles all cut sets on all layers.

   Figure 8.22 illustrates a generic system with a system split derived from a critical scenario determined by the developer. Each system is associated the generalized states "o.k." or "not o.k.". The later state is marked by the hatching. Four exemplary cut sets including according deduction paths from layer 2 are given. Each cut set consists of up to two independent faulty units. One cut set belongs to an externally supplied component that is not detailed to the lowest layer. To support processing of the derived data, each cut set has to be unique, and double fault based cut sets have to be pairwise disjoint with single fault based ones:

[27] "A cut set refers to the group of those elements or units which will make the system fail if their failure occurs. The minimum number of such units form the minimal cut set" (Verma and Ajit 2010, p. 85].)

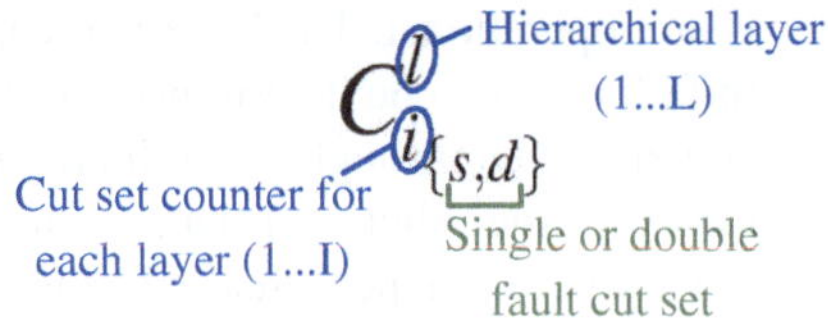

Fig. 8.21 Nomenclature for cut sets

$$C^l_{i_s} \cap C^g_{j_d} = \emptyset \quad \forall \quad \text{valid combinations of } i, j, l, g \qquad (8.2)$$
$$C^l_{i_{\{s/d\}}} \neq C^g_{j_{\{s/d\}}} \quad \forall \quad i \neq j \wedge l \neq g$$

Thereby, g and j denote values of the cut set counter and the hierarchical layer analogously to i and l. The proposed algorithm ensures disjoint cut sets by the structured segmentation of the system down to any hierarchical layer.

After completion of the above given two sub-steps of step 7, a list of cut sets exists for each failure scenario identified for the "result layer". According to the best knowledge of the developers performing the safety analysis, the combination of these cut sets then forms a minimal cut set for the overall system with the limitation that only up to two independent faults are regarded.

8.3.2.8 Step 8: Derive Failure Rate

Step 8 derives the failure rate of the overall system from the failure rates of components that are associated to identified cut sets. Therefore, two main tasks have to be addressed: At first, the probability of failure of the overall system due to one single or double fault based cut set is calculated. Therefore, both the mission time and the emergency operation interval of the vehicle are regarded. The calculations

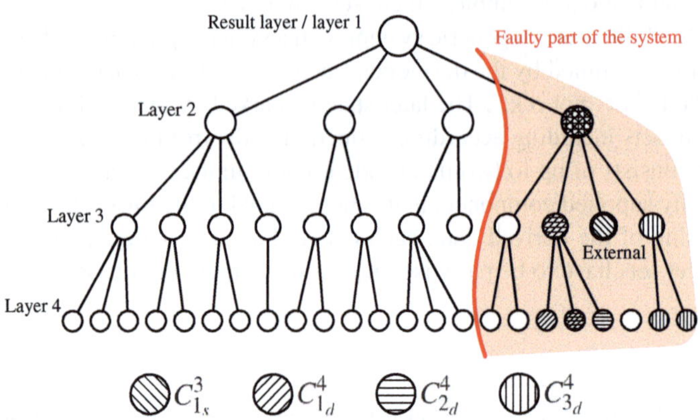

Fig. 8.22 Example cut sets for the faulty part of an example system

assume that aging of the vehicle is negligible during a single mission, and thus the failure rates are approximately constant. Secondly, the failure rates derived for the cut sets have to be combined throughout all hierarchical layers. This produces the overall failure rate of the vehicle. Iterative execution of the calculation in a time loop with modified failure rates regards aging effects. The following sub-steps result:

1. *System failure due to one single fault based cut set*: To start with, the failure rate of the overall system due to one single fault based cut set is determined. For a single fault based cut set, a failure of the system represents a single Markov transition. Accordingly, the failure probability per mission can be derived from the failure rate λ_i^l associated to the i-th single fault based cut set on layer l and the mission time T_M:
$$P(C_{i_s}^l) = 1 - e^{-\lambda_i^l \cdot T_M} \approx \lambda_i^l \cdot T_M \quad \text{for small lambdas.} \tag{8.3}$$

2. *System failure due to one double fault based cut set*: For calculation of the failure rates due to double faults, an approach presented by Sieglin (2009) is adopted. Sieglin (2009) derived a formula to calculate the failure probability of a power supply system consisting of two independent units with failure rates λ_1 and λ_2 and a diagnostic unit. The structure of this duplex system is given in Fig. 8.23. The probability p_{fail} of a system failure due to failure of both independent units can be calculated as:

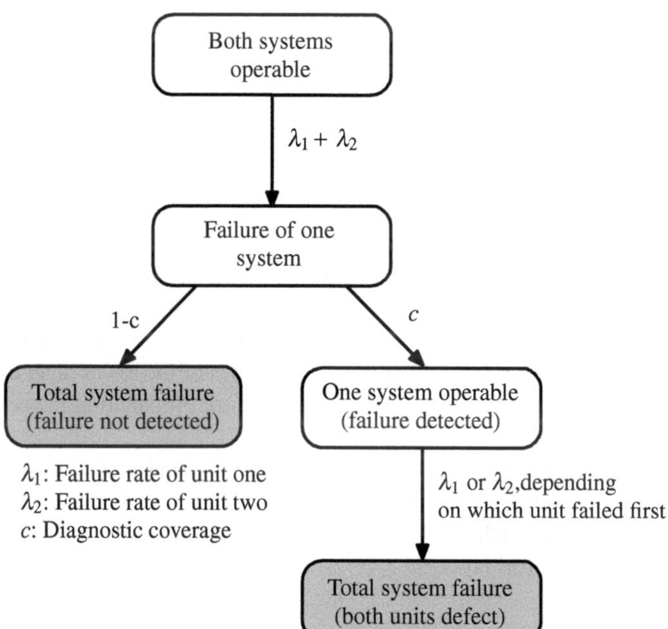

Fig. 8.23 State transitions in case of failure for a duplex system with diagnostic unit; figure similar to Sieglin (2009)

$$p_{\text{fail}} = 2\lambda_1\lambda_2(T_M T_{SS} - \frac{1}{2}T_{SS}^2), \tag{8.4}$$

with T_{SS} resembling the emergency operation interval. The formula can only be applied if three assumptions are valid: (a) At the start of a mission, all systems have to be in ok state. This can, e.g., be ensured by a detailed self check. (b) The exponential function $f(\Delta t) = 1 - e^{-\lambda \Delta t}$ can be approximated by the linear function $f(\Delta t) = \lambda \Delta t$ for small products of failure rates λ and time intervals Δt. And (c), the failure rates do not change depending on the order of occurrence of the two failures.

For the hierarchical approach the assumptions (a) and (b) are valid. Assumption (c) is mostly fulfilled due to the definition of independent virtual systems. If this assumption is not valid, the developer can easily introduce different failure rates for different orders of failures as the two failure scenarios with one failure occurring before the other one are treated independently during system analysis anyways. Thus, the formula can directly be used for most cases:

$$P(C_{i_d}^l) = 2\lambda_{i_1}^l \lambda_{i_2}^l (T_M T_{SS} - \frac{1}{2}T_{SS}^2). \tag{8.5}$$

Thereby, $\lambda_{i_1}^l$ and $\lambda_{i_2}^l$ resemble the failure rates of the first and second fault associated to the double fault based cut set with counter value i on the hierarchical layer l. The factor two ensures that both scenarios with one of the units failing first are already included. If for some failure combinations only one order of failure occurrence is possible or failure rates change for different orders, the evaluation algorithm accordingly neglects the factor 2 and treats both cases independently. Thus, the presented approach provides intrinsic means to handle timely effects. Other approaches have to make special extensions to handle these aspects as introduced by Mahmud et al. (2010) or Walker and Papadopoulos (2009) by means of temporal fault trees.

3. *Combination of all failure probabilities throughout all hierarchical layers*: To combine the failure probabilities throughout all hierarchical layers, a bottom up approach is followed. Starting with the lowest layer, the failure probabilities are combined. Then, the process is repeated on the next higher layer for the so far untreated cut sets. Thereby, the failure probabilities of the lower layer can simply be added due to independence of the cut sets before propagating the result to the next higher layer.

   The combined failure probability of multiple cut sets can be calculated as given in reliability engineering (Verma and Ajit 2010, p. 22). Transfered to the notion of the hierarchical approach, the following formula results for the failure probability within one hierarchical layer:

$$\begin{aligned}
P(C^l) = (+1)\cdot & \sum_{i=0}^{I} P(C^l_{i_{\{s/d\}}}) \\
(-1)\cdot & \sum_{i=0}^{i=I-1} \sum_{j=i+1}^{j=I} P(C^l_{i_{\{s/d\}}} \cap C^l_{j_{\{s/d\}}}) \quad \} \, \mathrm{OM}(\lambda^3) \approx 0 \\
(+1)\cdot & \sum_{i=0}^{i=I-2} \sum_{j=i+1}^{j=I-1} \sum_{k=j+1}^{k=I} P(C^l_{i_{\{s/d\}}} \cap C^l_{j_{\{s/d\}}} \cap C^l_{k_{\{s/d\}}}) \quad \} \, \mathrm{OM}(\lambda^4) \approx 0 \\
& \dots \\
(-1)^{I+1}\cdot & \, P(C^l_{1_{\{s/d\}}} \cap C^l_{2_{\{s/d\}}} \cap \dots \cap C^l_{I_{\{s/d\}}}) \quad \} \mathrm{OM}(\lambda^{I+1}) \approx 0.
\end{aligned} \tag{8.6}$$

Thereby, I denotes the number of cut sets within one hierarchical layer, j and k are helping variables defined based on i. OM$(\cdot)$ qualitatively resembles the order of magnitude of the computational terms in terms of a typical failure rate λ. The orders of magnitude are derived from Eqs. 8.3, 8.5, and the assumptions given in Eq. 8.2. Thus, cuts between a single and any double point fault based cut set do not exist. Resulting, only double fault based cut sets have to be regarded in line two and the following lines of Eq. 8.6. The cuts are calculated based on conditional probabilities resulting in the given order of magnitude in terms of λ. For example, the probability of a cut between two cut sets with identifiers i and j of the same hierarchical layer with one common fault within each cut set is determined as:

$$P(C^l_{i_d} \cap C^l_{j_d}) = P(C^l_{i_d}) \cdot P(C^l_{j_d} | C^l_{i_d}) \propto \lambda^l_{i_1} \cdot \lambda^l_{i_2} \cdot \lambda^l_{j_2} \quad \} \, \mathrm{OM}(\lambda^3) \tag{8.7}$$

Thereby, the faulty unit that is part of both cut sets is associated the failure rates $\lambda^l_{i_1}$ and $\lambda^l_{j_1}$, respectively. Based on the assumption of small lambdas, all terms with an OM higher than λ^2 are neglected.

Concluding step 8, the failure rates of the overall system per mission F results by bottom-up addition of all $P(C^l)$ derived for each hierarchical layer:

$$F = P(C) = \sum_{l=1}^{l=L} P(C^l). \tag{8.8}$$

Resulting, compliance with the given ASIL requirements can be verified.

8.3.2.9 Step 9: Derive Diagnostic Coverage

Finally, the diagnostic coverage (DC) on vehicle level can be estimated based on the results of the hierarchical approach. As defined for an FMEDA, the diagnostic coverage is calculated as (Löw et al. 2010):

$$DC = \frac{\lambda_{dd}}{\lambda_{dd} + \lambda_{du}}. \tag{8.9}$$

λ_{dd} and λ_{du} denote the cumulative probabilities of occurence of any dangerous failure that is detected (dd) or remains undetected (du).

This approach is now transfered to vehicle level. Thereby, it is important to note that the highly integrated system does not feature one separate diagnostic unit as, e.g., introduced by Sieglin (2009) and indicated in Fig. 8.23. The functionality of the diagnostic unit is distributed throughout the whole network. If furthermore functional redundancies between different actuators are regarded, inevitably the classical perception of the diagnostic unit has to be modified towards a globally operating system. The system may include complex knowledge about vehicle dynamics and the overall vehicle network.

The basic idea of the calculation is as follows: Due to targeted abstraction based on the generalized failure states, the hierarchical approach only regards dangerous faults. Furthermore, it is assumed that all safety critical functions are executed at least on redundant units that are insusceptible to common cause failures as otherwise the high safety requirements can not be met anyways. Thus, any functional unit can be seen as being structured similar to Fig. 8.23. Any single point failure leading to a failure of the overall system then must be caused by a failure of the diagnostic system and must contribute to λ_{du}. Resulting, the diagnostic coverage DC for a system that only has to tolerate one independent fault can be calculated according to Eq. 8.9 by relating faults that lead to an immediate failure of the system (λ_{du}) and faults that allow the vehicle to transition into its safe state (λ_{dd}).

The diagnostic coverage calculated as outlined then indicates the performance of the distributed diagnostic algorithms and signals whether the failure rate of the system is driven by high failure rates of individual units or missing quality of the diagnostic algorithms. On demand, automatic hints can be generated for which functions/units the performance of the distributed diagnostic algorithms is the lowest. Then, experts for the local system can derive according solutions.

8.3.3 Criticism of the Hierarchical Approach

This section provides a summarized criticism of the hierarchical approach to review the fields of application, novel contributions, limitations, and further usage of the generated data.

Fields of application: In modern vehicles, borders between individual safety critical functionalities as steering, braking and propulsion start to vanish. This becomes even more challenging if full-drive-by-wire vehicles are regarded. To limit additional costs for such systems due to numerous redundant hardware parts, integration of functionalities is unavoidable. Additionally, functional redundancies between different types of actuators have not yet been exploited for functional safety. Judging from the promising research results in the field of vehicle dynamics, these functional redundancies flanked by degradation concepts may hugely support cost-effective integration of highly safety critical EE systems into modern vehicles. Dedicated methods for safety analysis of such systems are missing. The hierarchical approach targets to close this rising gap.

Novel contribution: Summarized, some key aspects distinguish the hierarchical approach from other approaches:

- The hierarchical approach presents a *targeted way to system safety evaluation*. Thereby, front loading of knowledge about dependencies and critical states (virtual systems and generalized failure states) ensures that only relevant systems and faults have to be regarded in the later process. A goal-oriented split of the system in safety relevant components—away from the traditional domain oriented splits—is performed. Resulting, a high degree of abstraction can be achieved while still maintaining mathematical linkage for quantitative evaluation and proper documentation. Thus, especially the effort to derive quantitative failure rates can be reduced compared to other approaches (Papadopoulos et al. 2001).
- The hierarchical approach provides a tailored *structure function*[28] of the investigated system that neglects unnecessary components. The structure function can be visualized in different ways or be reused for further efficient analysis of the system as, e.g., shown by Adachi et al. (2011), Herath et al. (2007), Rehage et al. (2005), or Sinha (2011).
- The tool chain implemented in the project highlights components with the highest impact on failure rates but also *indicates the contribution to the safety concept* on each hierarchical layer. Thus, the approach supports system level thinking and encourages failure handling on all hierarchical layers.
- The hierarchical approach provides a failure rate for the overall system taking into account a configurable *emergency operation interval* and *failure rate estimates of dedicated software* functions. Additionally, the *diagnostic coverage* at vehicle level is approximated based on single and double fault based cut sets and the according probabilities of failure.
- Due to the dedicated re-partitioning of the system into safety relevant components, new safety concepts can become obvious that are not supported by domain oriented thinking. Analogously, *functional redundancies* can be exploited for safety evaluation. On top level, these redundancies can be integrated into the safety concept intuitively. On the basis of the resulting system structure, algorithms as presented by Herath et al. (2007) could be used for optimized allocation of failure rates or failure detection mechanisms within components with regard to overall system failure rate and costs.

Limitations: The hierarchical approach contributes when evaluating a system in terms of safety. Naturally, there are clear limitations of application.

1. As already pointed out, the hierarchical approach is neither a process model nor a method that primarily supports the development of a product. The hierarchical

[28] The structure function defines the "dependence of the system state on the state of its components" (Gertsbakh (2000), p.1).

approach supports evaluation of an already drafted vehicle architecture. Still, iterative application can support system development.

2. Additionally, the strong tailoring of the hierarchical approach on safety evaluation on the one side reduces work effort, but on the other side may also be unsuitable for extended investigation of a system, e.g., for a full investigation of reliability measures. If the hierarchical approach is extended to also investigate according scenarios, work effort approaches the one of already existing methods. The other way round, if generalization during the hierarchical approach is overdone to further decrease work effort, safety estimates may become more pessimistic. Still, results from common methods as FMEA support the developer to set the abstraction level appropriately.
3. The hierarchical approach is focused on quantitative evaluation of failure rates but does not evaluate fulfillment of process requirements opposed by ISO 26262. Process requirements derived from ASIL levels are a vital aspect for safety evaluation (Palin et al. 2011) and significantly contribute to development costs. The hierarchical approach could contribute to reduction of these local requirements by identification of local redundancies that could then allow ASIL decomposition. Still, handling of functional redundancies across different types of actuators in term of ASIL classification is a so far completely un-investigated topic in research and development.
4. Until now, the hierarchical approach focuses only on the basic vehicle control functions. Other functions provided by the human machine interface are not regarded. Still, this contribution holds the view that none of these aspects is relevant if the driver can no longer control the main actuators of the vehicle. Thus, the vehicle control system forms the basis for any other applications and should be treated separately. This perception is, e.g., backed by the generic safety life cycle for intelligent transport systems, especially Driver Assistance Systems, outlined by Carsten and Nilsson (2001).

Outlook: The information on system architecture and dependencies gathered by the hierarchical approach could be further exploited for online knowledge representation in the vehicle. Based on the known dependencies among systems and faults, the diagnostic heuristics could be (automatically) complemented to specifically take into account interactions on vehicle level. Multiple possible applications are thinkable that could make use of the well structured information on the system architecture from a safety point of view. Especially, systems dealing with self representation and online failure handling by degradation are possible fields of application and are investigated in the project MOBILE.

8.4 Safety Evaluation of MOBILE with the Hierarchical Approach

This section evaluates the safety of the vehicle control function of MOBILE using the hierarchical approach introduced in the previous section. As MOBILE is a primarily student driven university project, some restrictions have to be regarded:

- Up to 20 students were working on parts of the vehicle in parallel. Each student works on a specialized field on a low hierarchical level. The students are assumed to be the experts for a specific field. The work on higher levels is mostly done by members of the scientific staff.
- Failure rates are not available for all parts of MOBILE. For these parts typical values were derived from literature. Failure rates of software under development is roughly estimated based on previous in-field experience.
- Aging effects are approximated by typical bath-tub curves given in literature as, e.g., by (Reif 2009, p. 261) or (Verma and Ajit, 2010, p. 2). Thereby, one observes slightly higher failure rates of hardware components in early phases of the part's life time and a significant increase towards the end of the life time.
- Only the hierarchical layers "vehicle", "system" and "subsystem" have so far been taken into account (see also marks in Fig. 8.2). Some individual components of the underlying hierarchical layers are currently being investigated and results are fed back to the evaluation of MOBILE.
- Processes performed during development of MOBILE do not comply to the requirements imposed by ISO 26262.
- As the construction of MOBILE is not yet finished, only qualitative results are given that origin from quantitative but not yet complete data.

Still, the safety evaluation of MOBILE demonstrates the applicability and benefits of the hierarchical approach. Quantitative figures give a rough impression of the safety level. Additionally, relative changes in failure rates after modifications to components or the architecture of electronics can be observed. Working with a group of students showed that the hierarchical approach supports splitting of the complex vehicle design task into a number of smaller work packages that are easier to handle. At the same time, the system context is kept available and traceable for all developers.

8.4.1 Assumptions for the Safety Analysis of MOBILE

As indicated in Sect. 8.3, top level assumptions on the mode of operation of MOBILE are required to evaluate the functional safety of MOBILE: The *mission time* of MOBILE is limited to 30 min. After 30 min, the lead-acid drive batteries are assumed to be emptied anyways or the test driver is expected to have a break. In case of a failure, the emergency operation interval that has to be guaranteed is set to 30 s. This time span suffices to get MOBILE to a safe halt even if the failure occurred while driving at MOBILEs top speed of approx. 160 km/h (44 m/s). It is assumed that *only one independent fault* has to be tolerated. For MOBILE, one "point in time" is defined as a 4 ms time slot. This slot length is derived from the cycle time of the FlexRay network in MOBILE that facilitates synchronization of all network nodes and precise triggering of the diagnostic algorithms. If two faults occur within a 4 ms time slot, they are treated as a double fault at one point in time. A similar assumption for small diagnostic time intervals is, e.g., made by Sieglin (2009).

8.4.2 Evaluation of Complexity of the Hierarchical Approach

On "system layer" of MOBILE, eight units were defined: front and rear axle control system consisting of the according FTUs, user input control system (also embodied by the according FTU), two power supply systems, emergency off systems for front and rear drive motors and the stability control system (Table 8.5). Due to the design of MOBILE, these systems can be regarded as insusceptible to common cause failures—except loss of power. If cross couplings between the systems exist, the couplings are assumed to be irrelevant during the emergency operation interval of 30 s, e.g., low voltage buffer batteries can compensate the loss of charging power due to failure of the high voltage system. In particular, this independence of the elements at "system level" led to the definition of the virtual systems as given and not to the classical system partitioning into braking, drive and steering system. For each of the chosen virtual systems, 2 to 9 generalized failure states, not including the "ok"/"no failure present" state, were defined. Resulting, 31 failure states have to be evaluated on vehicle layer. Thereby, the controllability of the vehicle has to be evaluated after occurrence of a given first and second system failure, summing up to 702 state transitions. Especially, for the second faults, several transitions need not be regarded as they do not furthermore impact the controllability of the vehicle. Additionally, several transitions are identical for more than one system and thus only have to be considered once. For each failure scenario, the state of the vehicle is well defined as the generalized failure states are part of a first order Markov Chain. Thus, all relevant information is contained in the state descriptions and no knowledge about the failure history is needed. Given the knowledge about the effects of the system failures on vehicle dynamics, it takes the developer approximately an hour to go through all states and define the according consequences. Table 8.5 shows a simplified classification for MOBILE after the first failure for each system. Within the tool environment, the classification is done graphically based on Excel tables by color highlighting. For the evaluation of MOBILE on vehicle level, several experiments with a 1:5 scale vehicle were performed to estimate the effect of actuator or power supply failures on the controllability of the vehicle (Töpler 2010; Lieberam 2011; Goldschmidt 2012). Additionally research results of other groups were taken into account to fully exploit functional redundancies. Still, the classification at vehicle level is a challenging and not fully solved task from a scientific point of view but easy to handle formally, which allows the researcher to focus on his main tasks.

Table 8.5 Graphically assisted failure classification at vehicle level

system	generalized failure states/effect of first system failure				
FAC_Sys	destabilizing	neutral	ok		
RAC_Sys	destabilizing	neutral	ok		
EOffVA_Sys	defect off	defect on	ok		
EOffRA_Sys	defect off	defect on	ok		
ESup1_Sys	all off	12V off	48V off	HV off	... ok
ESup2_Sys	all off	12V off	48V off	HV off	... ok
SC_Sys	destabilizing	off	ok		
UI_Sys	defect	only loss of braking	only loss of steering	ok	

key:
FAC_Sys / RAC_Sys: front/rear axle control system
EOffVA_Sys / EOffRA_Sys: Emergency off system for front/rear axle
ESup1_Sys / ESup2_Sys: power supply 1/2
SC_Sys: stability control system
UI_Sys: user interfacing system
vehicle operable after failure of system: yes, no, no failure

The failure states on "system layer" are derived from approximately 60 failure states on "subsystem layer". In average, on system level approx. 100 state transitions have to be evaluated per system. Thereby, the behavior of the system for all "first faults" has to be considered. Additionally, selected "second faults" have to be investigated. Second faults that have to be regarded are identified automatically top down from "vehicle level". The number of state transitions that have to be investigated by the developer serves as an estimate for work load and complexity. If compared to "vehicle level" and depending on the individual system, the individual researcher on system level has to evaluate a similar amount of relevant combinations.

On lower levels (component and elementary) the number of total failure states furthermore increases but again can be handled due to the partitioning into virtual systems and allocation of tasks to local experts. Third party components can easily be integrated at any hierarchical level. Within the project MOBILE several such components exist (steering motors, drive motors, etc.).

As mentioned, the evaluation process in the project MOBILE is supported by an Excel Sheet. Necessary calculations and the linking between hierarchical layers are automatically derived from "graphical" inputs of the user (compare Table 8.5). As the input tables are continuously being updated during the development process, the current state of the vehicle with regard to safety as well as the most critical components are known at any point in time. The generalized failure states including proper documentation support transparency and long time usability of the results of the safety analysis. These state descriptions also form the basis for discussions between experts in different fields and on different hierarchical levels. A further extension of the tool environment to automatically link graphical architecture descriptions (fault

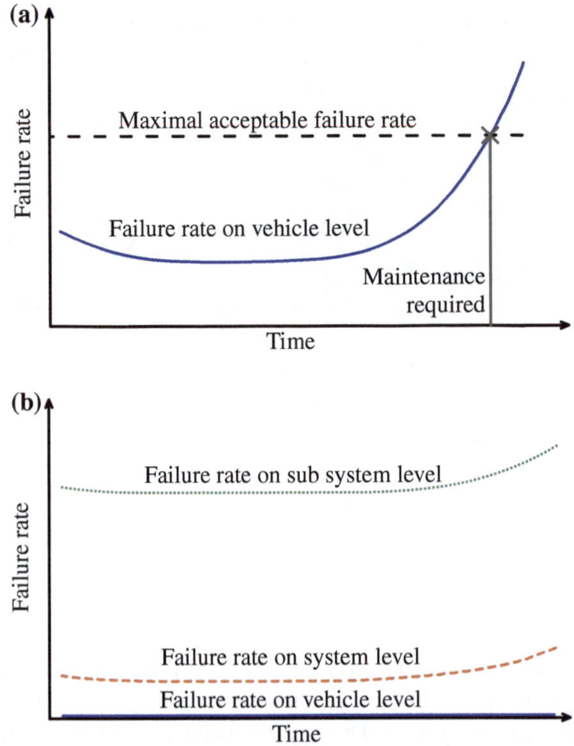

Fig. 8.24 Qualitative failure rates on "vehicle layer" (**a**) and for comparison on "vehicle", "system" and "subsystem layer" (**b**) over lifetime of the vehicle

trees, reliability block diagrams) or descriptions of state transitions (Markov chains) with the inputs in the Excel environment would be useful. Currently, these steps are performed manually, which is acceptable for the scope and scale of the project.

Summarized, the analysis results for MOBILE can serve as a well documented and tailored safety report and support continuous monitoring during development. The tailoring of the analysis by front loading knowledge on dependencies lowers work effort compared to other hierarchically structured approaches.

8.4.3 System Monitoring and Failure Rates

Figure 8.24a illustrates the failure rates of MOBILE at vehicle level over lifetime. Thereby, the failure rate was calculated using the approach detailed in Sect. 8.3 for several points in time. The curvy form of the graph with high increase in failure rates towards the end of the vehicle lifetime results from the assumed aging of hardware parts. As introduced in Sect. 8.3.2.4, software parts that feature a high probability of failure are also taken into account—differently from the approach in ISO 26262. Of course, these failure rates are highly volatile, but are several orders of magnitude higher than the failure rates of the underlying hardware and thus have to be considered. Of course, software components are unconcerned by aging.

Figure 8.24b visualizes the huge benefit for failure compensation in the vehicle by considering interactions at "system" and "vehicle layer". E.g., the curve for "system layer" considers only cross-compensations between different systems up to "subsystem level" and so on. On higher layers, these cross-compensations are more and more due to functional redundancies. Thus, a highly flexible vehicle as MOBILE especially profits. Analogously, the efficiency of the diagnostic coverage over lifetime is automatically derived from the gathered data.

Tendencies show, that the proposed integrated safety concept relying on functional redundancies can increase functional safety while also maximizing the functional benefit from additional actuators and limiting system costs due to reduction in required hardware redundancy. Still, final results can only be provided after the vehicle has been completed, and further experiments can be conducted.

8.4.4 Conclusion

This contribution introduces a novel system architecture for an experimental drive-by-wire vehicle with high functional integration and over-actuation. For this vehicle, a system architecture is derived top-down driven by according requirements. Especially, the top-down partitioning of the system can reveal novel structures also for series vehicles. Resulting, a system structure that exploits functional redundancies instead of hardware redundancies for safety purposes is presented. Exploiting functional redundancies necessitates a clearer definition of the safe state of the vehicle compared to typical part-oriented safe-state assumptions. For MOBILE, a model of the desired minimal vehicle dynamics is used. Consequently, control algorithms for vehicle dynamics play an important role in the proposed safety concept, and assessment of quality of these algorithms has to become more quantitative.

To evaluate the safety of complex and integrated systems as proposed for MOBILE, a hierarchical approach to safety analysis is introduced. The approach complements already existing means for safety evaluation by taking a holistic view of the overall vehicle. It especially focuses on the targeted evaluation of highly integrated systems that provide functional redundancies. Therefore, virtual systems and generalized failure states support early reduction of the number of faults that have to be analyzed quantitatively. At the same time, system partitioning promotes allocation of work packages to developers that are best suitable. As given, the proposed approach features some restrictions and potential for further development. Especially, questions related to a development process in industry as intellectual property, responsibilities, or process management are not regarded in this contribution.

Future work will focus on completion of the safety evaluation of MOBILE. Starting from there, the further usage of the structured information on the system architecture for online self-representation of the vehicle and diagnostics will be investigated. Another important topic of future work will be the ongoing evaluation of control algorithms for vehicle dynamics to exploit functional redundancies between different types of actuators by coordinated control of remaining actuators. In parallel, analysis of critical components of the EE system will go on with regard to functional safety and failure rates.

References

Abele, A.: Design and realization of an integrated safety concept based on an architecture model with the given example for the serial development of a powertrain control unit used in electric driven vehicle. In: Hybrid and Electric Vehicles, pp. 481–525. Braunschweig (2012)

Abele, M.: Modellierung und Bewertung hochzuverlässiger Energiebordnetz-Architekturen für sicherheitsrelevante Verbraucher in Kraftfahrzeugen. Ph.D. thesis, Universität Kassel, Kassel (2008)

Adachi, M., Papadopoulos, Y., Sharvia, S., Parker, D., Tohdo, T.: An approach to optimization of fault tolerant architectures using HiP-HOPS. Softw. Pract. Experience **41**(11), 1303–1327 (2011)

Anwar, S., Niu, W.: Analytical redundancy based predictive fault tolerant control of a steer-by-wire system using nonlinear observer. In: 2010 IEEE International Conference on Industrial Technology, pp. 477–482 (2010)

Arbitmann, M., Raste, T., Lauer, P., Kelling, E., Eckert, A., Rieth, P.E.: Motion Control—Zentraler Baustein zukünftiger funktional strukturierter Domänenarchitektur im Fahrzeug. In: AUTOREG 2011, pp. 375–387. Baden-Baden (2011)

Armbruster, M.: Eine fahrzeugübergreifende X-by-Wire Plattform zur Ausführung umfassender Fahr- und Assistenzfunktionen. Ph.D. thesis, Universität Stuttgart, München (2009)

Armbruster, M., Zimmer, E., Lehmann, M., Reichel, R., Sieglin, E., Spiegelberg, G., Sulzmann, A.: Affordable X-By-Wire technology based on an innovative scalable E/E platform-concept. In: IEEE 63rd Vehicular Technology Conference, pp. 3016–3020. Melbourne, Australia (2009)

Beal, C.E., Gerdes, J.C.: Experimental validation of a linear model predictive envelope controller in the presence of vehicle nonlinearities. In: 6th IFAC Symposium on Advances in Automotive Control. Munich (2010)

Bergmiller, P., Ibele, P., Maurer, M., Gerdes, J.C.: Development tool for dynamic drive control systems. ATZelektronik worldwide **2011–03**, 60–67 (2011)

Bergmiller, P., Maurer, M.: Flexible Versuchsträger als Testplattform für Antriebskonzepte in Elektrofahrzeugen. In: Schäfer, H. (ed.) 2012, Trends in der elektrischen Antriebstechnologie für Hybrid- und Elektrofahrzeuge, pp. 232–243. Expert Verlag, Renningen (2012)

Bergmiller, P., Maurer, M., Lichte, B.: Probabilistic Fault Detection and Handling Algorithm for Testing Stability Control Systems with a Drive-By-Wire Vehicle. In 2011 IEEE International Symposium on Intelligent Control (ISIC), pp. 601–606. Denver (CO), USA (2011b)

Bernard, M., Buckl, C., Döricht, V., Fehling, M., Fiege, L., von Grolmann, H., Ivandic, N., Janello, C., Klein, C., Kuhn, K.-J., Platzlaff, C., Riedl, B.C., Schätz, B., Stanek, C.: Abschlussbericht des vom Bundesministerium für Wirtschaft und Technologie geförderten Verbundvorhabens "eCar-IKT-Systemarchitektur für Elektromobilität". ForTISS GmbH, Garching (2010)

Bertacchini, A., Pavan, P., Tamagnini, L., Fergnani, L.: Control of brushless motor with hybrid redundancy for force feedback in steer-by-wire applications. In: 31st Annual Conference of IEEE Industrial Electronics Society, 2005. IECON 2005, pp. 1407–1412. Raleigh, USA (2005)

Blanc, S., Bonastre, A., Gil, P.: Dependability assessment of by-wire control systems using fault injection. J. Syst. Archit. **55**(2), 102–113 (2009)

Carsten, O.M.J., Nilsson, L.: Safety assessment of driver assistance systems. Eur. J. Transp. Infrastruct. Res. **1**(3), 225–243 (2001)
Collins: Collins English Dictionary 30th Anniversary Edition, 10th edn. William Collins Sons & Co. Ltd, London (2010)
Collinson, R.: Fly-by-wire. Comput. Control Eng. J. **10**(4), 141 (1999)
Cornelsen, K., Jänsch, D., Gerson, S., Nietschke, W., Maurer, M., Canders, W. R., Schumacher, W., Meyer, H.: InDrive Simulator—Innovative Tool for Simulating and Designing Complex Drive Structures in Real Operation. In: Hybrid and Electric Vehicles, pp. 166–186. Braunschweig (2011)
Dilger, E., Karrelmeyer, R., Straube, B.: Fault tolerant mechatronics [automotive applications]. In: 10th IEEE International On-Line Testing Symposium, pp. 214–218. IEEE Computer Society (2004)
Dominguez-garcia, A.D., Kassakian, J.G., Schindall, J.E.: A Backup System for Automotive Steer-by-Wire, Actuated by Selective Braking. In: 35th Annual IEEE Power Electronics Specialists Conference, pp. 383–388. Aachen (2004)
Euchler, M., Bonitz, T., Mitte, D., Geyer, M.: Bewertung der Fahrsicherheit eines Elektrofahrzeugs bei stationärer Kreisfahrt. ATZ - Automobiltechnische Zeitschrift **2010–03**, 206–213 (2010)
Freitag, G., Kuhn, K.-J.: Hochintegrierter Antrieb: Radnabenantrieb ohne Reibbremse. In: Schäfer, H. (ed.) Trends in der elektrischen Antriebstechnologie für Hybrid- und Elektrofahrzeuge, pp. 73–83. Expert Verlag, Renningen (2012)
Gadda, C.D., Laws, S.M., Gerdes, J.C.: Generating diagnostic residuals for steer-by-wire vehicles. IEEE Trans. Control Syst. Technol. **15**(3), 529–540 (2007)
Gertsbakh, I.: Reliability Theory With Applications to Preventive Maintenance. Springer, Berlin (2000)
Goldschmidt, D.: Entwicklung eines fahrdynamischen Stabilitätsprogramms für ein Drive-by-Wire-Versuchsfahrzeug. Diplomarbeit, TU Braunschweig (2012)
Hammerschall, U.: Flexible Methodenintegration in anpassbare Vorgehensmodelle. Technische Universität München, Dissertation (2008)
Hasan, M.S., Anwar, S.: Sliding mode observer based predictive fault diagnosis of a steer-by-wire system. In: Proceedings of the 17th International Federation of Automatic Control World Congress, pp. 8534–8539. Seoul, Korea (2008)
Hayama, R., Higashi, M., Kawahara, S., Nakano, S., Kumamoto, H.: Fault tolerant architecture of yaw moment management with steer-by-wire, active braking and driving-torque distribution integrated control. SAE Automotive Electronics Series, 2008–01-01 (2008)
He, L., Zong, C., Wang, C.: A steering-by-wire fault-tolerance control strategy based on multi-dimension gauss hidden Markov model. In: International Conference on Intelligent Control and Information Processing, pp. 227–230. Dalian, China (2010)
Heiner, G., Thurner, T.: Time-triggered architecture for safety-related distributed real-time systems in transportation systems. In: Symposium, Twenty-Eighth Annual International symposium on Fault-Tolerant Computing, pp. 402–432. IEEE Computer Society, Washington, DC (1998)
Herath, I., Roberts, C., Arvanitis, T.N., Bold, A.: Satisfying design constraints for automotive safety-critical systems. SAE Automotive Electronics Series, 2007–01-14 (2007)
Isermann, R., Beck, M.: Modellbasierte Methoden zur Erhöhung der Verfügbarkeit und Sicherheit von Fahrwerkkomponenten. AUTOREG 2011, pp. 679–690 (2011)
Isermann, R., Schwarz, R., Stölzl, S.: Fault-tolerant drive-by-wire systems. IEEE Control Syst. Mag. **22**(5), 64–81 (2002)
Johannessen, P.:SIRIUS, : Technical Report 01. Department of Computer Engineering Chalmers University of Technology. Göteborg, Sweden (2001)
Johannessen, P., Ahlström, K., Torin, J.: Conceptual design of distributed by-wire systems. SAE Automotive Electronics Series, 2002–01-02 (2002)
Johannessen, P., Törner, F., Torin, J.: Actuator based hazard analysis for safety critical systems. In: Computer Safely Reliability Security, vol. 3219, pp. 130–141 (2004)
Johannessen, P., Törner, F., Torin, J.: Experiences from model based development of drive-by-wire control systems. In: Kleinjohann, B., Gao, G.R., Kopetz, H., Kleinjohann, L., Rettberg, A. (eds.)

Design Methods and Applications for Distributed Embedded Systems, pp. 103–112. Springer, Boston (2004)
Kelling, N.A., Heck, W.: The BRAKE project—centralized versus distributed redundancy for brake-by-wire systems. SAE Automotive Electronics Series, 2002–01-02 (2002)
Kim, M.H., Lee, S., Lee, K.C.: Kalman predictive redundancy system for fault tolerance of safety-critical systems. IEEE Trans. Industr. Inf. **6**(1), 46–53 (2010)
Koehn, P., Eckrich, M., Smakman, H., Schaffert, A.: Integrated chassis management : introduction into BMW's approach to ICM. SAE Technical Paper Series 1(1219), (2006)
Köhler, R., Broy, J.: Markov-Ketten und Autokorrelation in der Sprach- und Textanalyse. In: Köhler, R., Broy, J. (ed.) Glottometrika 5 Bochum (1983)
Legler, H., Gehrke, B., Krawczyk, O., Schasse, U., Rammer, C., Leheyda, N., Sofka, W.: Die Bedeutung der Automobilindustrie für die deutsche Volkswirtschaft im europäischen Kontext (2009)
Lieberam, J.: Entwicklung eines Softwaresystems zur Zustandserfassung und -regelung im Kraftfahrzeug. Diplomarbeit, TU Braunschweig (2011)
Löw, P., Pabst, R., Petry, E.: Funktionale Sicherheit in der Praxis, 1st edn. Heidelberg: dpunkt.verlag GmbH (2010)
Mahmud, N., Papadopoulos, Y., Walker, M.: A translation of state machines to temporal fault trees. In: 2010 International Conference on Dependable Systems and Networks Workshops, pp. 45–51. Chicago, USA (2010)
Maier, M.W., Rechtin, E.: The Art of Systems Architecting, 3rd edn. CRC Press Taylor & Francis Group, Boca Raton (2009)
Masak, D.: Der Architekturreview. Springer, Berlin (2010)
Maurer, M.: Flexible Automatisierung von Straßenfahrzeugen mit Rechnersehen. Dissertation, Universität der Bundeswehr München, Düsseldorf (2000)
Maurer, M.: Automotive systems engineering—a personal perspective. In: Maurer, M., Winner, H. (eds.) Automotive Systems Engineering. Springer, Heidelberg (2013)
McLaughlin, S.B.: Analytic assessment of collision avoidance systems and driver dynamic performance in rear-end crashes and near-crashes. Ph.D. thesis, Virginia Polytechnic Institute and State University, USA (2007)
Mehmood, A., Easa, S.M.: Modeling reaction time in car-following behaviour based on human factors. Int. J. Appl. Sci. Eng. Techn. **5**(14), 93–101 (2009)
Miller, P.: A Prototype distributed architecture for safety critical automotive systems. SAE Automotive Electronics Series, 2007–01-16 (2007)
Mishra, P.K., Naik, S.M.: Distributed control system development for flexray-based systems. SAE Automotive Electronics Series, 2005–01-12 (2005)
Mitzlaff, M., Lang, M., Kapitza, R., Schröder-Preikschat, W.: A membership service for a distributed, embedded system based on a time-triggered flexray network. In: 2010 European Dependable Computing Conference, pp. 155–162. Valencia, Spain (2010)
Motruk, B., Diemer, J., Ernst, R., Buchty, R., Berekovic, M.: IDAMC : A many-core platform with run-time monitoring for mixed-criticality. In: 14th International High Assurance Systems Engineering Symposium Omaha, USA (2012)
Muenchhof, M., Beck, M., Isermann, R.: Fault-tolerant actuators and drives—structures, fault detection principles and applications. Ann. Rev. Control **33**(2), 136–148 (2009)
Müller, K., Steinbach, T., Korf, F., Schmidt, T.C.: A real-time ethernet prototype platform for automotive applications. In: 2011 IEEE International Conference on Consumer Electronics - Berlin (ICCE-Berlin), pp. 221–225. Berlin (2011)
Neudörfer, A.: Konstruieren sicherheitsgerechter Produkte. Springer, Heidelberg (2011)
Palin, R., Ward, D., Habli, I., Rivett, R.: ISO 26262 safety cases: compliance and assurance. In: 6th IET International Conference on System Safety, pp. 1–6. Birmingham, UK (2011)
Papadopoulos, Y., McDermid, J., Sasse, R., Heiner, G.: Analysis and synthesis of the behaviour of complex programmable electronic systems in conditions of failure. Reliab. Eng. Syst. Saf. **71**(3), 229–247 (2001)

Park, T.-j., Han, C.-s., Lee, S.-h.: Development of the electronic control unit for the rack-actuating steer-by-wire using the hardware-in-the-loop simulation system. Mechatronics **15**(8), 899–918 (2005)

Pfeffer, P., Harrer, M.: Lenkungshandbuch. Wiesbaden: Vieweg+Teubner Verlag | Springer Fachmedien Wiesbaden GmbH (2011)

Philipps, J.: Kontrolle ist gut, Misstrauen ist besser: Funktionale Sicherheit für integrierte Softwarefunktionen. In: Schäfer, H. (ed.) Trends in der elektrischen Antriebstechnologie für Hybrid- und Elektrofahrzeuge, pp. 129–140. Expert Verlag, Renningen (2012)

Pimentel, J.: Safety-reliability of distributed embedded system fault tolerant units. In: IECON'03. 29th Annual Conference of the IEEE Industrial Electronics Society, pp. 945–950. Roanoke, USA (2003)

Piyabongkarn, D., Lew, J.Y., Rajamani, R., Grogg, J.A., Yuan, Q.: On the use of torque-biasing systems for electronic stability control: limitations and possibilities. IEEE Trans. Control Syst. Technol. **15**(3), 581–589 (2007)

Pruckner, A., Stroph, R., Pfeffer, P.: Drive-By-Wire. In: Eskandarian, A. (ed.) Handbook of Intelligent Vehicles, pp. 235–282. Springer, London (2012)

Rausand, M., Hoyland, A.: System reliability theory—models, statistical methods and applications. Wiley, Hoboken (2009)

Rehage, D., Carl, U.B., Vahl, A.: Redundancy management of fault tolerant aircraft system architectures—reliability synthesis and analysis of degraded system states. Aerosp. Sci. Technol. **9**(4), 337–347 (2005)

Reichel, R., Armbruster, M.: X-by-Wire Plattform—Konzept und Auslegung. at—Automatisierungstechnik **59**(9), 583–596 (2011)

Reif, K.: Automobilelektronik, Eine Einführung für Ingenieure, 3rd edn. Wiesbaden: Vieweg+Teubner GWV Fachverlage GmbH (2009)

Reinold, P., Nachtigal, V., Trächtler, A.: An advanced electric vehicle for development and test of new vehicle-dynamics control strategies. In: 6th IFAC Symposium Advances in Automotive Control. Munich (2010)

Richter, D., Köhnen, A.: Sicherheitsziele für zukünftige Elektro-Fahrzeuge: Sicherheitsarchitektur für den elektrischen Antrieb basierend auf den Anforderungen der ISO 26262. In: Schäfer, H. (ed.) Trends in der elektrischen Antriebstechnologie für Hybrid- und Elektrofahrzeuge, pp. 95–100. Expert Verlag, Renningen (2012)

Rieth, P.E.: Das mechatronische Fahrwerk der Zukunft. In H. Winner, S. Hakuli, & G. Wolf (eds., 2012), Handbuch Fahrerassistenzsysteme, pp. 626–631. Vieweg+Teubner Verlag | Springer Fachmedien Wiesbaden GmbH, Wiesbaden (2012)

Rohe, M.: Entwicklung der Gesamtfahrzeugstrategie eines E-Fahrzeugprototyps mit Torque Vectoring. In: Schäfer, H. (ed.), Trends in der elektrischen Antriebstechnologie für Hybrid- und Elektrofahrzeuge, pp. 101–111. Expert Verlag, Renningen (2012)

Sakurai, K., Matsubara, M., Hoshino, M.: Membership middleware for dependable and cost-effective X-by-wire systems. SAE Automotive Electronics Series, 2008–01-04, 1–9 (2008)

Sangiovanni-Vincentelli, A.: Quo Vadis, SLD? reasoning about the trends and challenges of system Level design. Proc. IEEE **95**(3), 467–506 (2007)

Schäuffele, J., Zurawka, T.: Automotive Software Engineering—Grundlagen, Prozesse, Methoden und Werkzeuge. Friedr. Vieweg & Sohn Verlag/GWV Fachverlage GmbH, Wiesbaden (2004)

Schroer, R.: Flight control goes digital [Part Two, NASA at 50]. IEEE Aerosp. Electron. Syst. Mag. Part Two **23**(10), 23–28 (2008)

Schwall, M.L., Gerdes, J.C.: A probabilistic approach to residual processing for vehicle fault detection. In: Proceedings of the 2002 American Control Conference, vol. 3, pp. 2552–2557 (2002)

Siedersberger, K.-H.: Komponenten zur automatischen Fahrzeugführung in sehenden (semi-), autonomen Fahrzeugen. Dissertation, Universität der Bundeswehr München (2003)

Sieglin, E.: Beitrag zur Energieversorgung eines innovativen Drive-by-wire-Fahrzeugkonzepts. Dissertation, Technische Universität Dresden, Renningen (2009)

Sinha, P.: Architectural design and reliability analysis of a fail-operational brake-by-wire system from ISO 26262 perspectives. Reliab. Eng. Syst. Saf. **96**(10), 1349–1359 (2011)

Smakman, H., Köhn, I.P., Vieler, D.H.: Integrated Chassis Management—ein Ansatz zur Strukturierung der Fahrdynamikregelsysteme. In: 17. Aachener Kolloquium Fahrzeug- und Motorentechnik, pp. 1–13 (2008)

Starke, G.: Effektive Software-Architekturen. Carl Hanser Verlag, Munich (2008)

Sundar, M., Plunkett, D.: Brake-by-wire, motivation and engineering—GM sequel. SAE Automotive Electronics Series, 2006–01-31 (2006)

Tkachev, O.A.: Application of Markov chains for the reliability analysis of systems with a complex structure. Cybern. Syst. Anal. **19**(5), 96–101 (1983)

Töpler, S.: Entwicklung eines Abgleichreglers für die Fahrzeug Längs- und Querdynamik. Diplomarbeit, TU Braunschweig (2010)

Touloupis, E., Flint, J.A., Chouliaras, V.A., Ward, D.D.: A fault-tolerant processor core architecture for safety-critical automotive applications. SAE Automotive Electronics Series, 2005–01-03 (2005)

Trächtler, A., Niewels, F. Integrierte Querdynamikregelung mit ESP, AFS und aktiven Fahrwerksystemen. In: Isermann, R. (ed.) Fahrdynamik-Regelung, pp. 237–251. Friedr. Vieweg & Sohn Verlag I GWV Fachverlage GmbH, Wiesbaden (2006)

Tucci-Piergiovanni, S., Mraidha, C., Wozniak, E., Lanusse, A., Gerard, S.: A UML model-based approach for replication assessment of AUTOSAR safety-critical applications. In: IEEE 10th International Conference on Trust, Security and Privacy in Computing and Communications, pp. 1176–1187. Changsha, China (2011)

Verma, A.K., Ajit, S.: Reliability and Safety Engineering. Springer, London (2010)

von Vietinghoff, A.: Nichtlineare Regelung von Kraftfahrzeugen in querdynamisch kritischen Fahrsituationen. Dissertation, Universität Karlsruhe (2008)

Walker, M., Papadopoulos, Y.: Qualitative temporal analysis: towards a full implementation of the fault tree handbook. Control Eng. Pract. **17**(10), 1115–1125 (2009)

Waraus, D.: Steer-by-wire system based on flexray protocol. In: Applied Electronics, pp. 269–272. Czech Republic, Pilsen (2009)

Wilwert, C., Navet, N., Song, Y.Q., Simonot-Lion, F.: Design of automotive X-by-wire systems. In: Zurawski, R. (ed.) The Industrial Communication Technology Handbook, pp. (29–1)–(29–34). CRC Press, Boca Raton (2005)

X-by-Wire Project (1998). Brite-EuRam 111 Program. X-By-Wire—safety related fault tolerant systems in vehicles, final report

Zhen, B., Altemare, C., Anwar, S.: Fault tolerant steer-by-wire road wheel control system. In: Proceedings of the 2005 American Control Conference, pp. 1619–1624. Portland, USA (2005)

Zuo, G., Kumamoto, H., Nishihara, O., Hayama, R., Nakano, S.: Quantitative reliability analysis of different design alternatives for steer-by-wire system. Reliab. Eng. Syst. Saf. **89**(3), 241–247 (2005)

Part IV
Evaluation of Perception Capabilities

Chapter 9
Reference Systems for Environmental Perception

Mohamed Brahmi

9.1 Introduction and Motivation

The performance of advanced driver assistance systems (ADAS) is strongly dependant on the quality of the used environmental perception sensors and algorithms.

Therefore, the quality of the environmental perception needs to be assessed in order to allow a formative[1] and a summative[2] evaluation of the ADAS perception system.

This quantitative assessment of the perception sensors and algorithms requires the use of a reference system that can provide "ground-truth" information on the environment of a vehicle with higher accuracy and reliability than the used ADAS sensors. Besides accuracy and reliability, this reference system must satisfy other requirements, which will be discussed later.

Reference systems can be used at different stages during the development process of ADAS systems. Based on Maurer (2013), the systematic design of driver assistance systems can be illustrated through Fig. 9.1.

1. During the specification phase of new ADAS functions, the requirements on the environmental perception are derived from requirements on the function. At this stage, the use of a reference system can help verifying whether the intended sensor system satisfies these requirements. Hence, depending on the result of this verification, these specifications can be validated or changed without having

[1] Formative evaluation aims at the assessment of a system/process during the design and development. The assessment results obtained are used as feedback to improve this system/process.

[2] Summative evaluation deals with the assessment of the end performance of a system/process without feeding back the results.

M. Brahmi (✉)
Institute of Control Engineering, Technische Universität Braunschweig, Hans-Sommer-Str. 66, D-38106 Braunschweig, Germany
e-mail: brahmi@ifr.ing.tu-bs.de

M. Maurer and H. Winner (eds.), *Automotive Systems Engineering*,
DOI: 10.1007/978-3-642-36455-6_9, © Springer-Verlag Berlin Heidelberg 2013

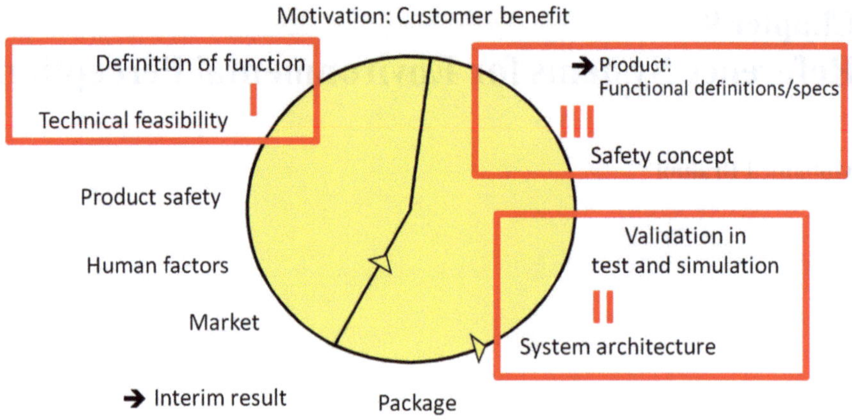

Fig. 9.1 Systematic design of driver assistance systems based on Maurer (2013)

developed the whole system yet. This helps to decrease development cost and time. In addition, the given specifications of the perception sensors can be verified.

2. For the development of perception algorithms, the detailed knowledge of the sensor behavior is crucial. Through sensor interpretation and inverse models, the detection of objects in the environment can be optimized.
   At a further stage, where simulation of the whole system or parts of it is needed, a forward sensor model is essential to simulate the real sensor. Combined with virtual environment simulation tools and real or prototyped Electronic Control Units (ECUs), the sensor model allows the test of a system in different forms (Hardware in the Loop, Software in the Loop, Vehicle in the Loop). The development of such simulation and interpretation models can be achieved by using reference systems that provide the "ground-truth" allowing an analysis of the sensor effects and properties such as separability, resolution and accuracy. These effects can then be artificially reproduced through these simulation models as shown in Brahmi (2010). In order to enhance the perception algorithms, these effects, when occurring online, can be correctly interpreted using the inverse sensor models. In addition, the internal object management in an object tracking system requires so called forward sensor models to predict new measurements from the actual state. Using a reference system, these sensor models can be improved and refined.
   All these measures, as part of the formative evaluation, can help optimizing the perception systems of ADAS.
3. At the end of the development process, final testing and benchmarking can be performed to assess the whole system in a summative evaluation.

9.2 Definitions

9.2.1 ADAS-Perception Terminology

- **Object**
  An object is a spatial element from the real world which can be described by a set of attribute information (geometry, pose ...).
  In perception systems for ADAS, an object can be a vehicle, bike or a pedestrian. Also abstract elements such as free spaces or road marking can be considered as objects.
- **Machine Perception**
  The perception system consists of sensors and algorithms that perform the measurement and the subsequent interpretation operations of the measurement results, in order to provide a representation of the surrounding environment.

9.2.2 Metrology Definitions and Their Meanings for ADAS

In this part, some important metrology definitions will be presented as well as their impact and meaning for the perception sensors and algorithms in ADAS.

- **Quantity**
  In JCGM (2012, p.2), the quantity is defined as a "Property of a phenomenon, body, or substance, where the property has a magnitude that can be expressed as a number and a reference".
  In ADAS-Perception these quantities are properties of relevant objects in the environment surrounding the ego vehicle (distance, velocity, acceleration, class, dimensions...).
- **Measurand**
  The Measurand is defined as "Quantity intended to be measured" according to JCGM (2012, p.17).
  The measurand is a particular quantity subject to measurement.
- **Measurement**
  "Set of operations having the object of determining a value of a quantity" (JCGM 2012, p.17).
  The measurement is a part of the perception operation and thus its quality affects the overall performance of the perception.
- **Repeatability (of results of measurements)**
  "Closeness of the agreement between the results of successive measurements of the same measurand carried out under the same conditions of measurement" (JCGM 2008, p.35).
  In ADAS perception systems, this is only given by static measurements where both ego-vehicle and target objects are not moving.

- **Reproducibility (of results of measurements)**
  The reproducibility, however, is defined as "Closeness of the agreement between the results of measurements of the same measurand carried out under changed conditions of measurement" in JCGM (2008, p.35).
  When the ego vehicle moves, the measurement cannot be replicated under the same conditions, but since the sensors are measuring the "same quantity" over time, these results can be considered as replicates of the measurements and can be analyzed to assess the accuracy of the sensor.
- **Reference value**
  JCGM (2012, p.53) defines a reference value as a "Quantity value used as a basis for comparison with values of quantities of the same kind".
  The reference value can be considered as the best estimate of the theoretical true value, which can only be obtained under perfect conditions.
- **Reference data**
  "Data related to a property of a phenomenon, body, or substance, or to a system of components of known composition or structure, obtained from an identified source, critically evaluated, and verified for accuracy" (JCGM 2012, p.53).
  The reference data is the measurement data collected from a reference system and is considered as the "ground-truth" data to compare sensor measurements to. Therefore, the reference system must be thoroughly verified.
- **Measurement error**
  In JCGM (2012, p.22), the measurement error is defined as "Measured quantity value minus a reference quantity value".
  In order to make a proper comparison between the reference and the sensor measurements, a correct spatio-temporal alignment is essential.
- **Measurement precision**
  According to JCGM (2012, p.22), the precision means the "Closeness of agreement between indications or measured quantity values obtained by replicate measurements on the same or similar objects under specified conditions".
  The environment of the vehicle to be perceived is made of static and dynamic objects. Only when the object and the vehicle are static, replicate measurements are given by repeatability; otherwise these replicates are given through reproducibility. The precision is related to the random part of the error. To assess the precision of the perception system, the difference between its measurements and those of the reference system is calculated at every time cycle. Afterwards, the spread of the distribution (2nd moment) of the measurement error can be used to assess precision.
- **Measurement trueness**:
  "Closeness of agreement between the average of an infinite number of replicate measured quantity values and a reference quantity value" (JCGM 2012, p.22).
  Measurement trueness is inversely related to systematic measurement error. It is usually expressed in terms of bias.
- **Uncertainty (of measurement)**
  As defined in JCGM (2012, p.25), the uncertainty of measurement can be described as a "non-negative parameter characterizing the dispersion of the quantity values being attributed to a measurand, based on the information used".

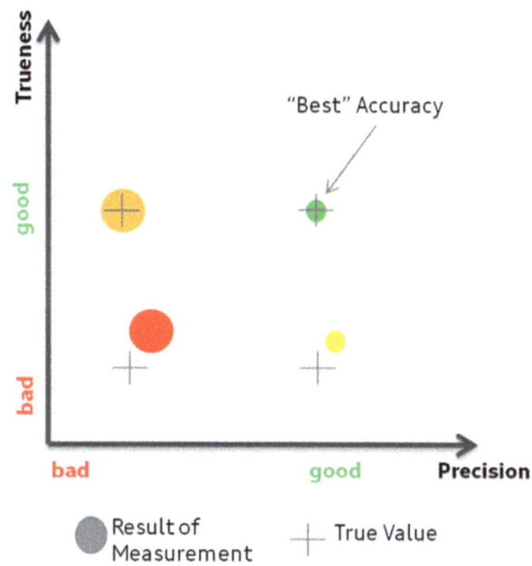

Fig. 9.2 Accuracy as the sum of trueness and precision

In statistics, dispersion means the extent to which values in a distribution differ from a fixed value. Since measurement values are merely estimates of the true value, it would be helpful to quantify the doubt that these values are within a given range. This quantification is expressed by the uncertainty of the measurement.

- **Measurement accuracy**
  "Closeness of agreement between a measured quantity value and a true quantity value of a measurand" (JCGM 2012, p.21).
  Accuracy refers to a combination of trueness and precision and can be used to describe the quality of a measurement. This is illustrated by Fig. 9.2.

9.3 Requirements on Reference Systems

Brahmi et al. (2012) have discussed the qualitative and quantitative requirements the reference system has to satisfy to provide reference data as defined above.

9.3.1 Qualitative Requirements

9.3.1.1 Mobility

Since the system under test is operating under different conditions and in different environments, the reference system also has to be usable in different conditions and environments. To be qualified as mobile, a system must be portable and self mobile.

- **Portability:** a portable system is moveable and should not be fixed to a predefined place.

- **Self mobility** is the system's ability to move: thereby the ability to accompany the system under test.

Therefore, the reference system must be mobile in order to cover different aspects when evaluating the perception system.

9.3.1.2 Self-Esteem

The system should be able to provide its own metrics of uncertainty. These can be associated with the measurement results to verify its quality. These metrics, such as the standard deviation, must furthermore be realistic, so that the system neither underestimates nor overestimates itself.

9.3.2 Quantitative Requirements

Besides qualitative requirements, the reference system must satisfy quantitative requirements.

9.3.2.1 Reliability

In order to obtain reference data, the reference system must be reliable. Accordingly, the reference data obtained from the system must guarantee a specific required quality during a given time of the measurement. This required time can be defined from the duration of a reference measurement where both reference and sensor data are collected.

9.3.2.2 Field of View

The reference system should be able to cover the entire Field of View (FOV) of the sensors under test even for sensor fusion configurations where the FOV is expanded.

The needed FOV to be covered by the reference system can then be derived from the FOVs of the system under test.

9.3.2.3 Accuracy

One of the most crucial aspects of a reference system is its accuracy of measurement. The requirements on the accuracy of the reference system can be derived from the required or expected accuracy of the sensor system. The reference system's accuracy should be higher than that of the system under test. The question: "How much better the reference system should be?" leads to a trade-off between the feasibility and required accuracy.

These accuracy requirements on the reference system can be derived in a direct manner if the ADAS function has accuracy requirements on directly measured quantities.

As discussed by Brahmi et al. (2012), some other functions have requirements on the accuracy of derived quantities which cannot be measured in a direct way. Through a measurement function in form of $\delta = f_{(}\varepsilon_1, \varepsilon_2, \ldots, \varepsilon_n)$, the derived quantity δ can be obtained from the directly measured quantities $\varepsilon_1, \varepsilon_2, \ldots, \varepsilon_n$. The uncertainty $\Delta\delta$ of the derived quantity can be accordingly obtained from the uncertainties $(\Delta\varepsilon_1, \ldots, \Delta\varepsilon_n)$ of the directly measured quantities by means of the linear law of error propagation: under the assumption that this quantities are uncorrelated

$$\Delta\delta = \frac{\partial f}{\partial \varepsilon_1}\Delta\varepsilon_1 + \cdots + \frac{\partial f}{\partial \varepsilon_n}\Delta\varepsilon_n$$

This method can be illustrated through the following example.

For several safety-relevant ADAS functions, such as collision warning/collision mitigation/collision avoidance (CM/CW/CA) described by Maurer (2012a), the quantity "Time to Collision" (TTC) is needed by the algorithms in order to take situation dependant decisions. TTC is defined as:

$$ttc := \frac{x_{rel}}{v_{rel}}$$

where x_{rel} denotes the relative longitudinal distance and v_{rel} the relative velocity between the ego vehicle and the target object. Applying the linear law of error propagation, we obtain:

$$\Delta ttc = \frac{1}{v_{rel}}\Delta x_{rel} - \frac{x_{rel}}{v_{rel}^2}\Delta v_{rel}$$

Taking into account the requirement $|\Delta ttc| \leq \Delta ttc_{\max}$ on the maximal tolerable error of the TTC defined by the safety-relevant ADAS function and taking (x_{rel}, v_{rel}) as operating point, the maximal tolerable errors Δv_{rel} and Δx_{rel} are graphically obtained as shown in the Fig. 9.3.

By taking (5 m, 5 m/s) as operating point and $\Delta ttc_{\max} = \pm 0.1$ s as maximal tolerable error, a relationship between Δx_{rel} and Δv_{rel} satisfying the requirement can be found. According to Winner et al. (2012), a TTC of 1s is assessed as critical by the drivers and thus (5 m, 5 m/s) can be taken as operating point. If v_{rel} can be measured within an uncertainty of ±0.4 m/s, then x_{rel} must be measured with ±0.1 m uncertainty and if $\Delta v_{rel} = \pm 0.3$ m/s, then Δx_{rel} must be ±0.2 m. The design trade-off mentioned above comes here into play.

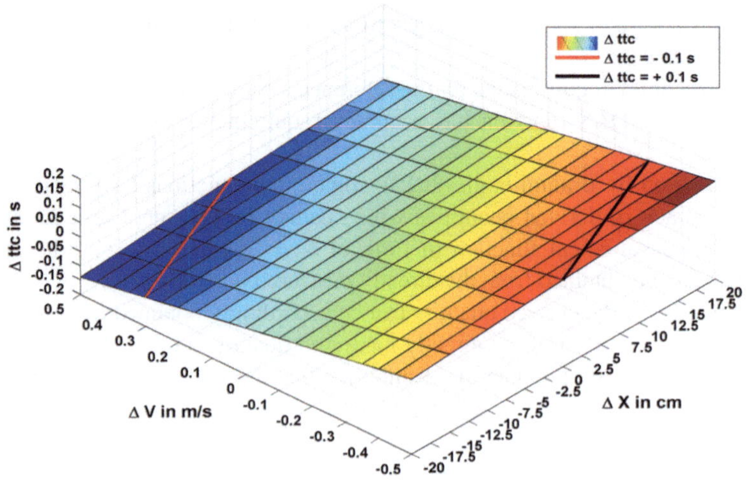

Fig. 9.3 Absolute Time to Collision estimation error depending on velocity and distance errors

9.3.2.4 Timing

- **Time delay**

  In order to be considered as ground truth, the reference data must be correctly time stamped. This can be achieved by satisfying following relationship:

$$|T_{meas} - \text{Timestamp}| \leq \varepsilon_{max}$$

  where T_{meas} denotes the real measurement time and ε_{max} the maximal tolerable time delay. Since every system is affected by latency problems, the timestamp of the reference data must take into account this latency by compensating it internally or providing it to the measurement system together with the reference data.

  Hence, the sensor measurement data can be correctly associated with the reference data for a proper comparison and evaluation.
- **Update rate**

  The reference system provides discrete measurements of the quantities in view. These measurements are mostly not synchronized with those of the sensor system and have therefore to be interpolated. This allows a reliable comparison between the reference data and the sensor data.

  Consequently, the measurement rate of the reference system must be sufficient to detect small changes of the quantity in view. Generally, the measurement frequency of the reference system must satisfy at least the Shannon-Nyquist Sampling theorem:

$$f_{Ref} \geq 2f_{Frzg}$$

where f_{Frzg} denotes the highest frequency component in the frequency spectrum of the desired vehicle signal as defined by Brahmi et al. (2012).

This condition is necessary but not sufficient to reconstruct the measured signal. Depending on the interpolation method between the samples, the signal can or cannot be correctly reconstructed. Generally, the interpolation error depends on both, the sampling rate and the used interpolation method. Using a linear interpolation method, it can be expected that the interpolation error decreases with an increasing sampling rate.

9.4 Example of a Reference System

The goal of environment perception for ADAS is to localize the ego-vehicle in relation to other objects within a given scene.

In order to generate ground-truth data of this localization, accurate and reliable localization systems are needed.

Due to the high accuracy and reliability requirements, automotive ADAS sensors cannot fulfill these demands.

However, perception sensors designed for special industry applications might be suited for the use as reference systems such as high definition laser scanners. These sensors are nevertheless not yet ready to be used as a reference system for ADAS perception, although they do achieve very accurate raw measurements. This is due to the difficulties of processing these raw measurements accurately to generate ground truth data.

Currently Differential GPS aided Inertial Navigation Systems (INS-DGPS) seem to be the most adequate solution to meet these requirements.

9.4.1 INS/DGPS

9.4.1.1 Overview

The standard GPS system cannot be used as a reference system because of its low accuracy (3–10 m). But due to differential correction DGPS techniques, the accuracy can be improved to reach 2 cm for Real Time Kinematic (RTK) applications according to Stephenson et al. (2011).

Despite of this enhancement of the accuracy, the DGPS system alone cannot be used as a reliable source of reference data. This is due to physical reasons, since sporadic errors may occur and GPS availability is dependent on constellation and surrounding buildings. Furthermore, the measurement rate is too low (1–20 Hz) to allow a reconstruction of sensor signals of dynamic objects. Therefore, DGPS is combined with Inertial Navigation Systems (INS) in order to overcome these problems and improve the overall system performance.

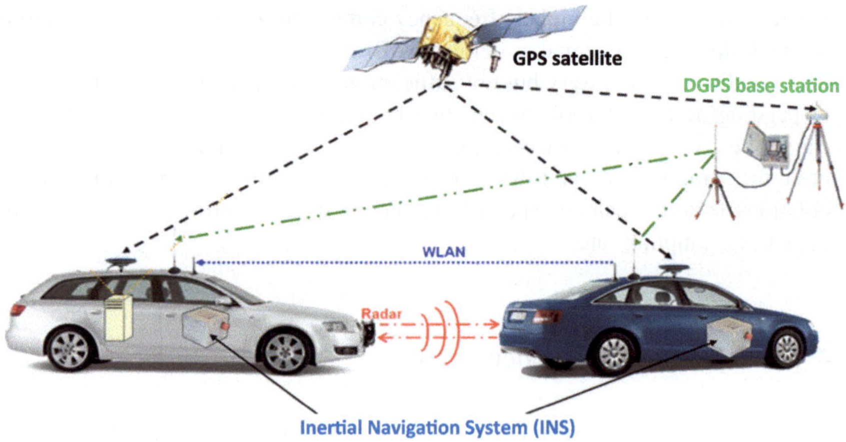

Fig. 9.4 Experimental set up of the reference system in Strasser et al. (2010)

Inertial Navigation Systems consist of accelerometers, gyroscopes on a common platform (IMU: Inertial Measurement Unit) as well as a navigation computer allowing the estimation of the position, velocity and orientation using the measured accelerations and rotation rates. Nonetheless, by reason of IMU measurement errors the INS diverges. By combining these relative measuring sensors with DGPS measurements of absolute positions, these drift errors can be periodically compensated.

9.4.1.2 Experimental Setup

An INS/DGPS system provides absolute measurements of the vehicle which is carrying it. ADAS sensors and perception algorithms generally provide relative measurements of the objects in the surrounding scene.

If many vehicles are equipped with an INS/DGPS system and if they can communicate with each other, a relative localization of the ego-vehicle relative to target-vehicles in the scene can be obtained (Fig. 9.4). This data can then be used as reference data for the ADAS sensors mounted on the ego-vehicle.

9.4.1.3 Satisfaction of the Requirements

The overall performance of these systems is strongly dependant on:

- **DGPS quality**: the DGPS' accuracy is the best accuracy the whole system can reach.
- **IMU quality**: the performance of the used accelerometers and gyroscopes also affects the computed results very strongly, since the different sensor's errors (bias

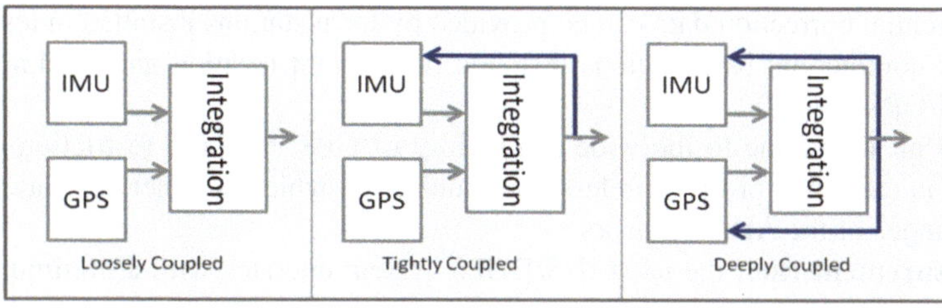

Fig. 9.5 Different integration strategies between IMU and GPS; IMU: Inertial Measurement Unit

Fig. 9.6 Time delay of the synchronization message

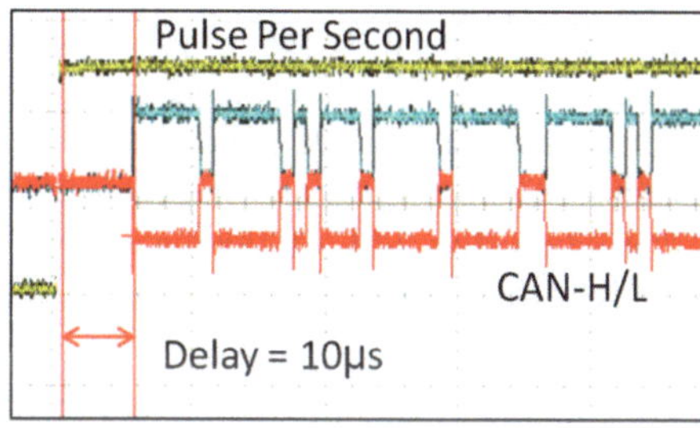

Table 9.1 Specified accuracy of the used INS/DGPS System in Scheyer and Von Hinüber (2012)

Measurement value	Accuracy
Absolute position	0.02 m
Relative position	0.03 m
	0.02 m
Velocity	0.01 m/s
Heading	0.05°
Roll/Pitch	0.01°

error, scale factor error etc.) result in position and velocity errors that grow over time.

- **DGPS/INS integration strategies**: depending on how the IMU part of the system is coupled with the DGPS part, the overall performance of the system can be improved. Generally, these strategies can be classified into the categories loosely, tightly and deeply coupled, as shown in Fig. 9.5. Schmidt and Phillips (2004) have shown that the performance of the system generally increases with the level of integration between the INS and the DGPS.

Since RTK DGPS techniques are becoming standards, the quality of the used IMU and the strategy of the integration between the IMU and DGPS parts of the INS/DGPS system determine the performance of the system (Fig. 9.6).

The used system has a deeply coupled integration and high class Fiber Optic Gyroscopes (FOG). Hence a good performance can be expected from this system.

The specification of the used INS/DGPS system is summarized in Table 9.1.

- **Mobility**: this system is portable and self mobile and can therefore be mounted in and moved by test vehicles carrying the perception system under test. The differential correction data can be provided by local stations installed in test sites or by commercial base stations installed in different location and sending over GSM/GPRS.

- **Field of View**: due to the wide availability of GPS, the only restriction of the FOV is the range of the Wireless communication which is generally wider than the ranges of the ADAS sensors.
- **Measurement rate**: the used INS/DGPS system operates with a minimum frequency of 100 Hz providing new measurement data every 10 ms. Due to the limited motion dynamic of passenger cars, this update rate satisfies by far the Shannon-Nyquist Sampling Theorem.
- **Latency**: the used reference system is able to deliver precisely time-stamped data using a synchronization mechanism based on the UTC (Universal Time Coordinate) and an accurate synchronization signal provided by the GPS Card. The latency of this signal and the corresponding synchronization message is lower than 11 μs (Fig. 9.6). All the measurements of this system are time-stamped relatively to this signal which is transmitted to the host system with a very low delay.

9.4.2 Validation of the Reference System

In order to be considered as a trustful source of reference data, the reference system should be validated by means of external and independent validation methods. These methods can be qualitative or quantitative.

Through analyzing the signals of the reference system, a "self contained" performance can be assessed by examining certain aspects of the signal such as its dynamic, range and stability. For an INS/DGPS the resulted position and velocity signals can be analyzed to detect non-realistic signal changes according to the dynamics of the vehicle. Additionally, the driven trajectory can be drawn over a geographical map allowing a rough statement on the quality of the system especially under poor GPS conditions.

The main issue of an INS/DGPS is its reliability, meaning the ability to guarantee a required performance among a given period of time. Hence another system with higher reliability can be used to validate this reference system under specific conditions.

Strasser et al. (2010) have compared several INS-DGPS systems to each other under the same conditions. The reproducibility and therefore the precision of these systems have been analyzed by driving repeatedly the same trajectory and with help of a high accuracy laser sensor mounted in the front of the vehicle. This laser sensor detects landmarks in form of building edges. By passing these landmarks, the edges are extracted and their positions in different rounds are analyzed. Using this verification method the self-esteem of these different systems can be examined by checking whether the output measurement uncertainty and the measurement quality are in agreement.

Table 9.2 Technical specifications of the used INS/DGPS system

	System 1	System 2	System 3
Gyro rate bias (deg/h)	6	0.75	36
Gyro rate scale factor (ppm)	1000	300	1000
Angular random walk (deg/$\sqrt{}$h)	0.12	0.16	–
Accel. bias (mg)	1	1	10
Accel. linearity and scale factor (ppm)	150	300	1000
Velocity random walk (μg/$\sqrt{}$Hz)	–	50	–

Table 9.3 Comparison results of System1 and 2 by Strasser et al. (2010)

System	Real precision (1σ)	Estimated precision (1σ)
System 1 (m)	0.75	0.43
System 2 (m)	0.43	0.19–0.5

Table 9.4 Specifications of a high performance IMU in Kennedy et al. (2006)

	Absolute Ref-IMU
Gyro rate bias (deg/h)	0.0035
Gyro rate scale factor (ppm)	5
Angular random walk (deg/$\sqrt{}$ h)	0.0025
Accel. bias (mg)	0.03
Accel. linearity and scale factor (ppm)	100

Table 9.5 RMS errors of System2 in Kennedy et al. (2006)

		System2
Position difference RMS (m)	North	0.038
	East	0.034
	Height	0.033
Velocity difference RMS (m/s)	North	0.007
	East	0.008
	Height	0.005
Attitude difference RMS (deg)	Roll	0.011
	Pitch	0.014
	Yaw	0.038

Table 9.2 summarizes the specifications of different INS/DGPS systems which have been compared to each other by Strasser et al. (2010).

The comparison results in Table 9.3 show that "System2" has a precise reproducibility and a good self-esteem. In contrast, "System1" tends to overestimate its quality which can be critical if used as reference system.

In Kennedy et al. (2006) a very high performance IMU, described in Table 9.4, was used in combination with post processing and smoothing methods so that the overall system (IMU + DGPS + Algorithms) can be considered as ground-truth (Tables 9.4 and 9.5).

The System2, using the same DGPS, is then compared to this ground-truth and the results are summarized in Table 9.5.

Table 9.6 Technical specifications of the LRR3-Sensor in Bosch (2009)

Range/Accuracy	0.5 . . . 250 m/ ± 10 cm
Field of View Horiz./Vert.	30°/5°
Velocity Range/Accuracy	−75 . . . + 60/ ± 0.12 m/s
Modulation	FMCW
Frequency range	76 ...77 GHz

Table 9.7 Evaluation results of the LRR3 Sensor in Brahmi (2010)

Quantity	Mean	Standard deviation
Lateral position (m)	−0.03	0.3
Longitudinal position (m)	0.3	0.015
Relative velocity (m/s)	0.1	0.05

These validation methods have shown that the used INS-DGPS system exceeds other similar systems.

9.4.3 Use-Case: Referencing an FMCW Radar Sensor

The Long Range Radar Sensor (LRR3) is used for ACC and other ADAS safety-relevant functions. It is based on FMCW (Frequency Modulated Continuous Wave) principals and allows the direct measurement of relative distance and velocity and the estimation of the lateral position of the objects in its FOV.

The specification of this sensor, as described in Table 9.6, can be evaluated and verified with the help of the INS-DGPS reference system.

Therefore, the ego-vehicle is equipped with the INS-DGPS system and the LRR3 sensor measuring the relative position and motion of a target vehicle which itself has to be equipped with the same INS-DGPS system. Thus, sensor and reference data can be collected and analyzed.

For a proper comparison, the reference and measurement data must be temporally and spatially aligned. Therefore, it has to be transformed into a common coordinate system and reference data must be interpolated between two reference measurement timestamps encountering the sensor measurement timestamp according to Brahmi (2010).

The results of the evaluation of the lateral and longitudinal position as well as the relative velocity are summarized in Table 9.7.

9.4.3.1 Lateral Position

While the ego and target vehicles are driving in a convoi, the measurements of the relative lateral position of a target vehicle are analyzed and compared to reference data. The statistical analysis of the measurement error allows making statements about the precision and the trueness of these measurements and thereby about its accuracy.

The precision of the measurements is given by the obtained standard deviation of 30 cm. This implies a wide distribution measuring the lateral distance, which can be explained by the limitations of estimation techniques of the lateral position.

However, the mean of the measurement is close to zero implying a good trueness.

9.4.3.2 Longitudinal Position

Similar to the previous measurement, the offset and standard deviation of the distribution of the error are calculated.

The obtained standard deviation of 1.5 cm implies precise measurement of the longitudinal distance. This can be explained with the fact that the sensor measures this quantity accurately with help of the detected frequency shift.

The distribution shows nevertheless an offset of 30 cm, which can be erroneously interpreted as deterministic error of the sensor.

This offset can be explained by the fact that the radar sensor measures where the reflectance is higher. In most cases, the rear of a vehicle is made of plastic based materials. Accordingly, the parts with more metal content inside the vehicle body are detected and measured by the sensor. This knowledge should be taken into account when developing safety-relevant functions based on this radar sensor to properly estimate the nearest point from a vehicle in front.

9.4.3.3 Relative Velocity

As expected, the measurement of the relative velocity is very accurate due to the use of the Doppler-effect and sophisticated signal processing methods. The found relative velocity measurement error is in the specified range of ± 0.12 m/s.

9.4.3.4 Time Delay

Thanks to its low latency and high accuracy, this reference system can be used to analyze the latency of the LRR3 sensor.

Generally, the latency results in a shift in some measured non constant quantities. Hence, the measurement and analysis of this shift can lead to the measurement of the latency of the sensor.

To measure the latency correctly, the use of an appropriate comparison quantity which can be well measured by the sensor is useful, in order to exclude other error sources. The relative longitudinal distance is used as a comparison quantity for analyzing the latency of the sensor. For this purpose, the ego follows a target vehicle so that the relative distance is varied in a sinusoidal form.

Taking into account the accuracy of the longitudinal distance measurement discussed above and by means of cross correlation methods applied to the sensor signal y_n and reference signal x_n after a time discretization and alignment with a sampling period T, the latency can be calculated according to Zhang and Xiaolin (2006) as following:

$$R_{xy}(m) = \frac{1}{N}\sum_{n=0}^{N-1} x_n y_{n+m}$$

where N denotes the length of samples and $m = -(N-1)\ldots 0 \ldots (N-1)$.

The result is a $2N-1$ vector R_{xy} which takes its maximum value at position n_0. The sensor delay can then be estimated to $\tau = n_0 * T$. For the LRR_3 sensor, the time delay has been evaluated to 120 ± 10 ms by using a sampling period $T = 10$ ms.

9.5 Conclusion

Reference systems allow the evaluation of the performance of ADAS perception sensors and algorithms and thus their optimization and improvement.

Therefore, the used reference system must meet some requirements in order to be utilizable as a verified source of reference data.

The comparison of the results allows the evaluation of the performance by means of statistical metrics. Nonetheless, there is a need to develop further metrics to make this evaluation more meaningful and exhaustive.

The presented INS-DGPS based reference system can be used to analyze the perception sensors and algorithms dealing with only one relevant target object. However, current and future ADAS functions need the detection of several relevant objects in the surrounding scene. For referencing a scene with multiple objects using this reference system, each dynamic object should be equipped with this system which increases the cost and efforts needed for the referencing.

Therefore an additional reference system should be developed to deal with these issues, allowing multiple-object referencing in combination with the described INS-DGPS.

Candidates for this solution could be a high definition laser scanner based system with a large field of view and all-round vision.

References

Bosch, R.: LRR3 3rd Generation Long-Range Radar Sensor. Robert Bosch GmbH, Germany (2009)

Brahmi, M.: Diploma thesis: Entwicklung. Implementierung und Validierung eines Sensormodells für zukünftige Radarsensorik. Technische Universität München, Germany (2010)

Brahmi, M., Hübner, M., Siedersberger, K.-H., Wegener, M.: Anforderung an ein Referenzsystem, ATZ Elektronik (2012)

JCGM: JCGM 100:2008 Evaluation of measurement data—Guide to the expression of uncertainty in measurement. The International Bureau of Weights and Measures (2008)

JCGM: JCGM 200:2012 International vocabulary of metrology—Basic and general concepts and associated terms (VIM), 3rd edn. The International Bureau of Weights and Measures (2012)
Kennedy, S., Hamilton, J., Martell, H.: GPS/INS Integration with the iMAR-FSAS IMU. XXIII FIG Congress Munich, Germany (2006)
Maurer, M.: Forward Collision Warning and Avoidance. In: Eskandarian, A. (eds.) Handbook of Intelligent Vehicles. Springer, London (2012a)
Maurer, M.: Automotive Systems Engineering—A Personal Perspective. In: Maurer, M., Winner, H. (eds.) Automotive Systems Engineering. Springer, Heidelberg (2013)
Scheyer, H., Von Hinüber, E.: ADAS Testing. Real-world measurements, Vision Zero International Magazine (2012)
Strasser, B., Bubb, H., Maurer, M., Siedersberger, K.-H., Siegel, A.: *Vernetzung von Test- und Simulationsmethoden für die Entwicklung von Fahrerassistenzsystemen (FAS)*. In: Tagung Aktive Sicherheit durch Fahrerassistenz, München (2010)
Stephenson, S., Meng, X., Moore, T., Baxendale, A., Edwards, T.: *Precision of Network Real Time Kinematic Positioning for Intelligent Transport Systems* (Nottingham Geospatial Institute, University of Nottingham, Nottingham, 2011)
Zhang, L., Xiaolin, W.: On the Application of Cross Correlation Function to Subsample Discrete Time Delay Estimation. McMaster University, Hamilton (2006)
Schmidt, G., Phillips, R.: INS/GPS Integration Architecture Performance Comparisons. The Charles Stark Draper Laboratory, Cambridge (2004)
Winner, H., Hakuli, S., Wolf, G.: Handbuch Fahrerassistenzsysteme. Komponenten und Systeme für aktive Sicherheit und Komfort (Vieweg+Teubner Verlag/GWV Fachverlage GmbH Wiesbaden, Grundlagen, 2012)

Chapter 10
A System Architecture for Heterogeneous Signal Data Fusion, Integrity Monitoring and Estimation of Signal Quality

Nico Dziubek

10.1 Introduction

10.1.1 Motivation

A large number of today's automobiles, right down to the compact car segment, are equipped with vehicle dynamics control and driver assistance systems. In general, each one of these functions has been developed with its own dedicated set of sensors, and applied independently of other sensors installed in the same vehicle. As a result, redundant measurements are performed, the advantages of which are currently only utilized in a few cases. In light of the ever-increasing networking of functions, the differing measurement principles of the individual sensors must be borne in mind so that discrepancies or inconsistencies arising from different measuring errors can be resolved. In addition, the availability, sampling rates, resolution and delay times of the measurement signals generally differ and depend on external conditions. This is due to different types of sensors and function principles.

The increasing powerfulness of microprocessors and the availability of bus systems in vehicles offer a basis for a central processing of the large quantity of data already available. An architecture created in this way for holistic processing of the measurement signals to cover all available data sources makes it possible to generate a consistent data record with increased accuracy. The quality of the data generated is evaluated based on this. This evaluation provides the user functions with additional information for further processing.

Particularly in the case of safety-critical systems, the outlay for detecting measurement and sensor errors is very high. Centralized evaluation of the signal integrity offers the potential of relieving the functions of a considerable part of error detection, and of improving error detection through use of the redundancies (Table 10.1).

N. Dziubek (✉)
Institute of Automotive Engineering, Technische Universität Darmstadt, Petersenstraße 30, D-64287 Darmstadt, Germany
e-mail: dziubek@fzd.tu-darmstadt.de

M. Maurer and H. Winner (eds.), *Automotive Systems Engineering*,
DOI: 10.1007/978-3-642-36455-6_10, © Springer-Verlag Berlin Heidelberg 2013

10.1.2 Goals

The structure of a system architecture for centralized, consistent fusion of random pieces of data is shown and evaluated using an example. The requirements to be met by a fusion filter for processing data from random types of sensors are determined, and implemented for a set of sensors by way of example.

The sensors used here are acceleration and yaw rate sensors produced using micro-electro-mechanical system (MEMS) technology combined to an inertial measurement unit (IMU) with 3 degrees of freedom, a single-channel (L1) GPS receiver that issues raw data (pseudo ranges and carrier phase measurement), as well as odometry sensors measuring angle pulses from all four wheels and the steering wheel angle.

In addition, an evaluation of the signal quality based on usage of redundancies is shown. For this purpose, a description of the signal quality by means of integrity and accuracy is shown, and components for such a description are shown by way of example.

10.2 State of the Art: Automotive Signal Fusion and Integrity

10.2.1 Overview

Approaches to sensor signal fusion and integrity monitoring in the recent past have been mainly made in the fields of aviation (Bickford et al. 1997), robotics (Soika 1997), railway (Liu et al. 2011) and power plants (Ibargüengoytia et al. 2001). In the automotive field of application, first approaches have been made to specify the requirements of integrity and accuracy. Feng and Ochieng (2007, p. 4), suggest the following Required Navigation Performance (RNP) for road vehicles:

Table 10.1 Required navigation performance for road vehicles (Feng and Ochieng 2007, p. 4)

Mode of operation	Horizontal accuracy (95 %) (m)	Integrity			Continuity (risk)	Availability
		Integrity risk	Horizontal alert limit (m)	Time to alert (s)		
Fleet management	25–1500	10^{-6}/hr	62–3750	15	10^{-5}/hr	0.997
Vehicle command and control	30–50	10^{-6}/hr	75–125	10	10^{-5}/hr	0.997
Automatic vehicle monitoring	30	10^{-6}/hr	75	10	10^{-5}/hr	0.997
Route guidance	5–20	10^{-6}/hr	7.5–50	10	10^{-5}/hr	0.997
Collision avoidance	1	10^{-7}/case	2.5	1	10^{-5}/hr	0.997

Many system designs dealing with automotive sensor fusion were developed with the aim of autonomous vehicle control. The most important methods used there will be shortly described in Sect. 10.2.2; they describe several ways for evaluating sensor and signal integrity, as well as methods for error detection and isolation.

This article will show appropriate criteria for selecting and combining these existing methods from a system design and integration point of view, i.e. which integrity and error detection algorithm fits the requirements for working with signals from cost-effective series sensors and offering useful results for the application together with driving dynamics control and driver assistance systems. Also, the ease of series application and usability in existing system architectures will be considered.

10.2.2 General Methods

In general, measurement and sensor errors are detected by comparing sensor signals with a reference. This leads to the conclusion that some kind of redundancy is necessary. The following redundancy methods shown by Pourret et al. (2008) are appropriate for sensor validation:

1. Hardware redundancy (the same physical property is measured by two or more identical sensors)
2. Analytical redundancy (mathematical conversion of other measured physical properties to the required one)
3. Temporal redundancy (statistical evaluation of multiple repetitions of the same measurement, therefore not appropriate for real-time applications)
4. Knowledge-based redundancy (using knowledge about the system considered for error detection)

An example given by Goebel and Agogino (1999) for knowledge-based redundancy is Fuzzy Logic, which is commonly used in power plants. It evaluates the validity of a sensor signal by its own history and does not need a mathematical description.

Another method for error detection is modelling statistical signal dependencies in a Bayesian Network, where a probability of validity is assigned to every sensor signal.

Further stochastic methods are the Normalized Innovation Squared (NIS)-Test, which is suitable for innovation-based fusion filters like the Kalman filter, and the Parity Space Method, which creates a matrix which is orthogonal to the observation (measurement) matrix. From both methods, a description of the relative measurement inconsistencies is derived and used for error detection.

10.2.3 Integrity in Navigation Applications

GNSS are a widespread source of the absolute position in navigation applications. Therefore, many concepts for evaluating integrity in GNSS-based navigation solutions are based on Receiver Autonomous Integrity Monitoring (RAIM).

In general, RAIM algorithms can be classified into Range Domain Methods, which are based on testing the GNSS raw measurements for inconsistencies, and Position Domain Methods, which are testing the resulting position solution for implausibility. A further classification can be done in the time domain, as there are Snapshot methods which take only the most recent measurement epoch under consideration, and sequential methods, which also take the signal history into account. It was shown by Brown (1992) that three classic snapshot RAIM Methods, namely Least Squares Residual, Parity Space and Range Comparison are mathematically identical. Young and McGraw (2002) added a normalized variant of the Solution Separation method to these equivalent schemes. Bhatti and Ochieng (2009) showed a method for detecting Slowly Growing Errors and Multiple Failures, which are known weaknesses of many RAIM algorithms. Feng and Ochieng (2007, p. 4), have defined the term Vehicle Autonomous Integrity Monitoring (VAIM) and suggest a RAIM algorithm adapted to road vehicles. Examples for sequential methods are given in Le Marchand et al. (2009), where the signal history is regarded in a Trajectory Monitoring Algorithm, and Toledo-Moredo et al. (2007), who use an Interactive Multiple Model Filtering.

10.2.4 Conclusion: State of the Art

In general, many different, proven methods for evaluating the integrity of signals or a sensor fusion system exist. Although first definitions of, and requirements for integrity in the automotive sector exist, no appropriate algorithm concept, and no system architecture has yet been defined, and the existing methods only partly fulfill the requirements. Hence, in this article a top-down approach will be shown, beginning at the requirements of automotive integrity, and leading to the definition of an automotive integrity and accuracy benchmark, as well as a system architecture suited for the integration into existing and future system setups.

An example of a sensor fusion system fulfilling the defined requirements will be shown. Still, in general, the selection of which algorithm, or which combination of algorithms, depends on the individual system architecture requirements and the type of fusion filter in use.

10.3 System Approach: Fusion of Heavily Heterogeneous Signals

10.3.1 Requirements

As a prerequisite for creating the system architecture, it is assumed that a vehicle constitutes a multi-sensor environment in which the individual signal sources do not necessarily communicate with each other; but such a scenario is not ruled out either.

The basis for fusion of measurement signals is the presence of redundancies, i.e. the measurement of identical physical values on different paths. It is of no relevance here whether this occurs through multiple, direct measurement of the same variable (hardware redundancy), conversion of a different measurement variable into the required variable (analytical redundancy) or through generation of virtual reference measurement values by means of model assumptions (knowledge-based redundancy). In addition, the fusion is to be implemented in such a way that, assuming an appropriate system and error modelling, the quality of the fused data improves through every inclusion of additional measurement signals. A further requirement arises from this, namely to strive toward using as many signal sources as possible with redundancies with other signal sources. Due to many and due to heterogeneous measurement principles of the measurement and signal processing technology used, considerable differences arise in the measured data, and these must be taken into consideration by the fusion filter:

- Data are recorded synchronously or asynchronously with other signal sources
- Different measurement resolution
- Different, possibly non-constant sampling rates
- Changing availability of information sources over time
- Dependencies on ambient conditions
- Dynamically changing accuracy during operation

Processing redundant pieces of data into a uniform, consistent result leads to non-specific effects on the overall data record output if errors in the measurements occur. This makes it difficult, if not impossible, for a user function to detect and treat errors. As a result, the fusion filter must forward information about errors and signal quality to the user functions. In any case, due to central processing of all signals involved in the fusion, integration of error detection and, if possible, compensation or isolation, in the fusion filter is self-evident. Compared to conventional error detection methods restricted to individual functions, improved detectability and treatment options for errors can be achieved here, as, in addition to the self-diagnosis ability of signal sources and model-based plausibilization methods available in most cases, it is possible to perform checks with further redundant signals.

In order to create only an improvement rather than a restriction of the data made available for user functions, it is a requirement for the data issued by the fusion that the data rate and resolution meet the requirements of the most demanding function in the system and their information content and degrees of freedom must be equal to or higher than the data used by the user functions prior to the introduction of the data fusion. There is a further requirement: due to the additional processing step introduced in the signal processing chain by the fusion, the resulting delay or group delay must at least be known and be as constant and as small as possible, the filter must have as little influence as possible on the bandwidth of the signals and these influences must be well known and must be issued as a further description of the signal quality. In particular, this also means that usage of signals with a large delay time must not lead to an increase in the latency of the fusion filter.

An appropriate architecture is designed below, a fusion filter devised and components of a quality description for the fused signals shown.

10.3.2 Architecture

From a structural perspective, data fusion is a new level located between the already present signal sources and the applications in the vehicle that use the data. The architecture has been already described in Dziubek et al. (2012), p. 3ff (Fig. 10.1).

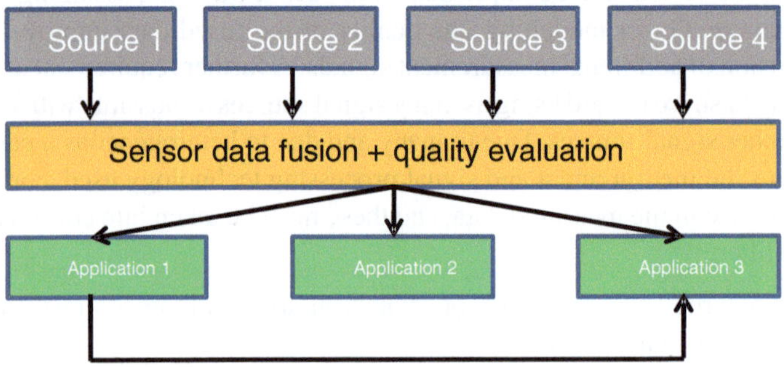

Fig. 10.1 Architecture of the system (Dziubek et al. 2012, p. 3)

This architecture causes decoupling of the applications from the signal sources that generate the data. In addition to the actual measurement values, the signal sources also provide information about their system states such as the results of a cyclic self-test. All these pieces of data are gathered together in the fusion block, and processed as a synchronous data record (measurement epoch).

Data fusion acts as a central virtual signal and information source for the applications. As data fusion provides the data for all applications centrally, consistency between the pieces of data is ensured. Communication between individual applications is thus made easier.

The signal sources do not have to be available simultaneously here. The pieces of data available for a processing step are always fused sequentially, as described in Sect. 10.4.2. Thanks to this asynchronous structure, different sampling rates of the measurement signals, fluctuating processing and run times of the signals or drop-outs of signal sources (e.g. GPS in the tunnel) do not need special treatment. A further advantage arising from this architecture is the fact that additional signals can easily be added. However, it must be borne in mind here that the time required for processing and sending data to the fusion algorithm is calculated for the purpose of consistent utilization of the measured values in the fusion filter. The more accurate the individual actual times of measurement are for every signal source for the purpose of correct chronological attachment of the data, the smaller the error caused by delays will be.

The diagnosis data issued by the signal sources are also collected and evaluated centrally. Data marked as invalid by the signal source itself are not included in the fusion. The quality of the data is also evaluated. As shown in Sect. 10.5.2, pieces of data are checked for plausibility among each other using stochastic methods.

This comparison is advantageous for the probability and speed of the detection of implausible measured values, compared to the self-analysis of the individual signal sources alone, as all system data, including model-based data, are available to the central algorithm in the form of test variables. This results in a clear statement regarding the integrity of the data.

As described in Sect. 10.5.3, the evaluation of the accuracy is performed on the basis of uncertainty regarding the output variables in the fusion filter as well as on the basis of general assumptions regarding quality descriptions of measurement variables. In a generalized approach, a set of parameters is defined on the basis of typical descriptions of a sensor data sheet. This set contains these characteristic features for describing the signal quality. These parameters are delivered by the signal sources in the form of additional items of information about the measurement variables. When the signal processing chain is being run through, they are updated every time a processing step is run through so that a description of all signal characteristics required for further processing is available to applications using the data, even if dynamic changes arise in the characteristics of the processing chain.

10.4 Fusion Filter

10.4.1 Concept

An example, according to widely according to Dziubek et al. (2011, p. 4ff), is given for a fusion filter approach that meets the central data processing requirements and can be used in the system architecture described before. The following data sources are used here:

- Inertial measurement unit (IMU): MEMS acceleration sensors and MEMS yaw rate sensors, each with three degrees of freedom in the form of a 3D basic system
- GNSS receiver: GPS single-channel (L1) receiver. Zogg (2009) describes how pseudoranges (distances between satellites and receiving antenna) and the differentiated carrier phase measurements (speeds between satellites and receiving antenna) are used; these are issued by the receiver in the form of raw measurement data.
- Odometry measurements: Wheel turning angle pulses with direction detection (wheelticks) of every single wheel as well as the steering wheel angle, which is converted into a wheel steering angle by means of the known static steering ratio.
- Model-based correction data: These pieces of data contain assumptions via checking of basic conditions (e.g. standstill) as well as specific assumptions for application in the automobile.

Data from external sources is not used for the purpose of correction or improvements.

10.4.2 Structure

The IMU is used as a basic system of data fusion as the variables measured and corrected by it describe a 3D movement in its entirety, it has the highest sampling rate of all data sources used in the system and is subject to no external availability restrictions, i.e. availability has been specified by the failure probability of this assembly unit. As described in Wendel (2007, p. 45), the IMU is combined with a strapdown algorithm. This component calculates the dynamic variables in the ground-fixed coordinate system on the basis of the vehicle-fixed yaw rates and accelerations measured by the IMU. Accelerations, yaw rates, speeds and the current position are issued. Measured acceleration by gravity, earth yaw rate and Coriolis acceleration are compensated for, and sensor errors calculated by the fusion filter are balanced out. This consists of an error state space extended Kalman filter similar to the one concepted by Wendel (2007, p. 194), together with sequential updating as described in Bar-Shalom et al. (2001, p. 88), which is executed at the same time the IMU is sampled. This evaluates the errors in the basic system "IMU + strapdown algorithm"; apart from this correction, no intervention is made in the calculation of the driving dynamic data. Cf. Fig. 10.3. As a result, the filter's inherent dynamics only have a minor influence on the signals issued by the strapdown algorithm, and a low, almost constant signal delay arises. The familiar phenomenon in which Kalman filters react with a delay to dynamic changes following a long stationary drive as the internal model relies too heavily on predictions is thus avoided; the modeled signal errors and their changes are largely independent of the current driving dynamics.

The filter calculates the modeled errors of the IMU-offset and scale factor errors-and of the navigation data via a linearized system model (extended Kalman filter) by using correction data. All correction data available at the time of execution are attached, taking their different, asynchronous sampling times into consideration. This is facilitated by the sequential update. After the Kalman filter's linearization and prediction step has been completed, the presence of new pieces of data is checked (individually for each data channel), and these are used for correction purposes if necessary. The superposition principle applies for the individual measurements due to the linearized system model-the filter's result after completion of the update sequence is identical to that of attaching the updates in a single step. As a result, an automated, dynamic adjustment to changing availabilities and sampling rates of correction data is achieved without changes to the filter's measurement or system matrices (Fig. 10.2).

As shown in Fig. 10.3, the fusion filter is divided into two general parts. The basic system performs the navigation calculation in the strapdown algorithm with IMU data and corrections calculated by the filter. In the correction system, the redundant data are offset against each other in a measurement model and the error volume determined. Errors in the signals used for calculating the corrections are also identified and corrected in the relevant measurement model. In this way, using the advantages of different measurement principles largely compensates for their disadvantages and overall accuracy is improved.

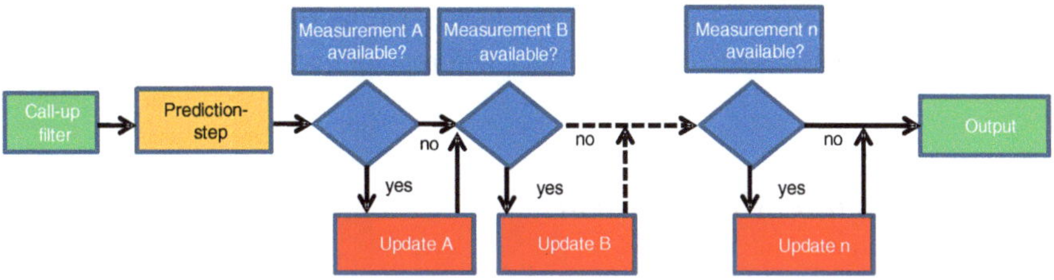

Fig. 10.2 Block diagram of sequential update (Dziubek et al. 2011, p. 5)

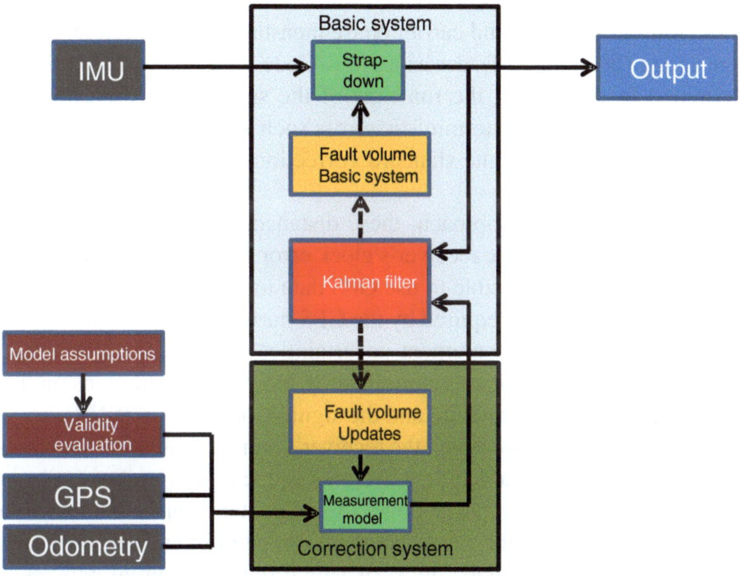

Fig. 10.3 Structure of the fusion filter (Dziubek et al. 2012, p. 7); IMU: Inertial Measurement Unit

Both parts of the filter have been created in 3D coordinate systems without any restrictions. Assumptions due to fitting the filter to the application in the automobile that might restrict the universality of the approach are not made here. As a result, all correction data used must also be available in three dimensions for the purpose of compatibility and conversion. If no pieces of data are available for individual degrees of freedom, e.g. for the speed along the vertical axis in odometry measurements, model assumptions are made for these degrees of freedom.

Model assumptions that are valid for two-lane vehicles in particular are attached as a correction measurement in the form of an additional virtual sensor. The validity of the models is specified on the basis of measurable basic conditions; if the validity criteria have been met, the calculated model data are attached. Thanks to this structure, the filter can also be adapted to other vehicle types by simply replacing a virtual sensor, while the basic structure of the filter will remain unchanged.

10.4.3 Correction Measurements

10.4.3.1 GPS Tightly Coupling

GPS raw measurement data are used in a Tightly Coupling by means of measurements for the purpose of error evaluation. As shown in Wendel (2007, p. 194ff), Code measurements from the L1 band and carrier phase measurements are fed into the fusion filter here. Zogg (2009, p. 14ff), describes how Code measurements (pseudoranges) are determined by measuring the run time of the signal from the satellite to the receiver. In order to correct deterministic errors such as atmospheric influences and clock errors in the satellites, the standard corrections documented in IS-GPS-200 (2006) are attached.

In the Tightly Coupling approach, these distance measurements are processed directly in the fusion filter. The receiver's clock error is also evaluated in the fusion. In this configuration, it is possible to use GPS data to improve the filter states, even if less than the four satellites required by the GPS during solo operation can be seen.

In order to obtain the measurement uncertainties required for processing these data in the Kalman filter, they are determined by means of calculations on the basis of the elevation angle of the satellite and the signal to noise ratio (SNR).

Depending on the type of receiver, the receiver's carrier phase measurements have a resolution of a fraction of a degree. Given a wave length of approx. 19 cm in L1, a distance resolution with accuracy down to millimeters can be achieved, but the measurement ambiguity of the range between the satellite and receiver cannot be resolved without special methods, most of which require external data, as can be seen in Mansfeld (2004, p. 156).

However, the distance covered between two times of measurement in the direction of the satellite position vector can be determined, presupposing interruption-free reception between two measurement epochs. As the times of measurement at the start and end of the distance measurement are known with a high resolution, the vehicle speed in this direction can be determined on the basis of this.

The weighting of the differentiated carrier phases for the Kalman filter is performed in the same way as for the pseudo range measurements.

10.4.3.2 Odometry Measurements

For odometric measurements (distance and heading measurements), the pulses from the wheel angle encoders (wheelticks) of the four individual wheels and the steering wheel angle are available. Unlike the IMU and GPS measurements, these measurements do not cover all six spatial degrees of freedom of movement.

In principle, the following three movement variables can be calculated by evaluating the wheel rotational speeds and the differences between them:

- 2D level speed / distance of the vehicle in a random point of the vehicle, assuming sideslip-free and slip-free driving
- Angular rate / rotation angle around the vehicle's vertical axis (yaw rate / yaw angle)

Two even movement variables are calculated for each wheel via a linear tire model using the steering angle and the active wheel forces. This means that eight measurement variables are available, the system is thus over-determined to the degree of five.

To resolve this, a least squares error (LSE) as described by Leinen (2010, p. 55ff), approach is used. This approach contains a linearized system model that resolves the over-determination and calculates the variances in the output variables directly and independently of external measurement variables.

As the Kalman filter's system model works with six degrees of freedom in 3D space, and the measurements therefore require the same dimension, but the odometry only covers three degrees of freedom, the remaining three degrees of freedom are calculated by means of model assumptions. Vehicle-typical basic conditions are incorporated here.

10.4.3.3 Virtual Sensors

The fusion filter core works without any restrictions by model assumptions. These, if necessary, are incorporated into the fusion filter as normal filter updates, and processed the same way as real sensor measurements. Therefore, these models are called virtual sensors. However, they are calculated completely on the basis of model assumptions and basic conditions valid for the vehicle. All variables required for determining the validity of these basic conditions, and hence for evaluating the validity of the model in question are known from measurements, and the data output are provided for all degrees of freedom required by the filter by means of calculations or assumptions. In this way, the filter that has been kept universally valid is provided with vehicle-typical basic conditions adapted to the actual application. These conditions can be adapted to other vehicles or vehicle types by replacing the virtual sensors.

10.5 Integrity and Accuracy

The fusion filter architecture described in Sect. 10.3.2, creates a basis that allows the integration of data that have been measured or determined by means of a model, and improves accuracy. The data output by this virtual sensor has varying qualities that depend on which data sources are currently available. In order to use the described advantages of the improvements achieved by the fusion, a method of describing these qualities is being investigated, widely according to Dziubek et al. (2012, p. 8ff). The goal of this integrity/accuracy indication approach is to define and calculate the qualities of signals on the basis of parameters, thereby facilitating further processing in any application, even a safety-critical one, while keeping the approach universal and independent of the concrete structure of the fusion filter.

10.5.1 Basis

The terms used (integrity and accuracy) are derived from the desire to create a general quality description for signals. The goal here is to obtain a description of the signal characteristics that is as easy and complete as possible in order to provide the functions using the data with all information about them required for further processing or control. The description of the signal characteristics begins in the data sources that assign the relevant quality information known from the technical specification to their output. These pieces of information are passed on along the entire signal processing chain, and adapted dynamically as signal processing takes place, so that a user function always receives the quality description corresponding to the actual measurement and processing. The tasks of such a quality description, which is also suitable for safety-critical systems, are as follows:

- Description of the state of the signal processing chain in the form of a general statement about the consistency of all available pieces of data which can be checked against each other.
- Provision of a description of the accuracy and errors of the measurement signals across the entire signal processing chain, from the data sources propagated right up to the data sinks.
- Dynamic calculation of the quality description in order to correctly model changes in the processing chain if these arise in adaptive filters, e.g. due to temporary non-availability of data sources or changes of the filter characteristics over time.
- Prediction of the maximum signal uncertainty over time that applies in the worst-case scenario. This, in particular, forms the basis for advanced driver assistance systems that only offer functions if the required minimum accuracy is available throughout the function's maximum operational time.

Measurement signals contain disturbances and errors with different causes. As these vary heavily in their effects and characteristics, a subdivision into different error classes is necessary in order to describe the signal quality. An examination of the signal on its own does not allow clear classification of these disturbances in the time or frequency domain. For example, a bias drift can be seen as a quasi-static constant or a low-frequency noise depending on the perspective. In order to achieve a clear subdivision, the effects are therefore distinguished on the basis of the physical creation and clarification principles as described by Wendel (2007, p. 70ff), and Kammeyer and Kroschel (1998, p. 9ff, p. 77ff, p. 157ff):

- Noise: Band-limited, zero-mean statistical distribution of the measurement signals around the expected value. Noise does not depend on the operating point, and approaches the expected value of zero given infinitely long averaging.
- Offset/bias: A value that is independent of the operating point and assumed to be constant if there are no changes in the ambient conditions. It is superimposed on the measurement results as an addition.
- Scale factor error: A value that depends on the operating point and assumed to be constant if there are no changes in the ambient conditions. It is multiplied by the measured value after the offset has been subtracted.
- Offset/scale factor drift: Changes to the offset or scale factor error over time that are due to changing environmental conditions, e.g. temperature changes, fluctuations in the supply voltage.
- Ageing of the sensor: Is neglected as such effects apply over significantly longer time spans compared to the uninterrupted switch-on duration of the sensors.
- Timing uncertainty: Effects that influence the signal timing properties, e.g. delay times or filter group delays.
- Frequency influences: Changes in the signal that can be described in the frequency range, e.g. pass-band and stop-band of a filter.

10.5.2 Integrity

10.5.2.1 General Definition of an Integrity Statement

The possible states of the result of checking a signal's integrity are as follows:

- An error has been detected.
- No error has been detected within defined limits.

Evaluation of a signal's integrity is performed by means of a comparison with references. Examples of these references are redundant measurements of the same variable, calculated on the basis of other measurements, or provided by limit values or model assumptions. Likewise, the data output by the fusion is to be incorporated so that possible errors in the fusion filter can be detected in order to evaluate the integrity of the entire system. Detection of an error presupposes that the error can be observed and that a reference value that contradicts the observations is available. As these conditions depend on external influences and the system's operating point, a stochastic description of the probability of detection, the time until detection and the confidence of the integrity statement are necessary. A definition and an approach for describing the confidence is given in Sect. 10.5.2.2. The integrity statement can therefore not be used as a basis for assuming that the system is completely free of errors (necessary condition: observable contradiction). By contrast, a detected contradiction in the measurement signals provides a reliable statement regarding an error condition of the system (sufficient condition: a contradiction has actually been observed). A further source for error detection comes in the form of the status data output by the self-diagnosis of the data sources; these are also used to evaluate the integrity.

10.5.2.2 Confidence of the Integrity

Measurement signals exhibit a dispersion. The maximum permissible dispersion has generally been specified by the data source specification. An integrity statement will only be regarded as positive if deviations of the same measurement value calculated along different paths, when compared with other measurement signals, also lie within these limits. When measurement signals are compared with each other, the evaluation of the overlapping of the individual dispersion bandwidths gives rise to a benchmark for the consistency of the signals, and hence forms a basis for evaluating the integrity. The confidence with which the integrity statement is made has not yet been taken into consideration here. Two characteristic criteria are suggested as a means of describing the confidence of the integrity. They are described below.

Clearly, an integrity statement based on comparing measurement signals with a narrow dispersion bandwidth is more accurate than one with a broad bandwidth. In the same way, a minor difference between measurement signals is to be evaluated more positively than a major one, which may lie at the very edge of the specification. In light of this, a relative benchmark that evaluates the overlapping of signals must be used to describe the confidence. This is called the consistency benchmark below.

In systems with data sources that are not always available (e.g. GPS) or with dynamically changing signal characteristics (e.g. in the data output by the fusion), the relative evaluation of the signals in relation to each other can mean that the occurrence of non-availability of a source that is very accurate compared to the other data sources leads to a rising confidence. This is the case if a difference between the signals disappears due to the occurrence of non-availability of a source with a low dispersion bandwidth. However, since the entire confidence level has deteriorated here, this situation must be described with an additional absolute benchmark that evaluates the maximum achievable confidence in the event of an ideal overlapping of all available measurements. This is called the dispersion benchmark below.

10.5.2.2.1 Consistency Benchmark / Relative Confidence

The goal of this part of the confidence benchmark is to describe the consistency of the available measurement signals in relative terms. To this end, the consistency of the measurements as well as their uncertainty is weighted against each other. The requirements for the consistency benchmark are derived from these criteria:

- Independent of the number of compared measurement signals and their absolute stochastic uncertainty
- Higher relative weighting of measurement signals with low uncertainty

- Independent of the absolute values of the signals / operating points, but dependent on the differences between the signals
- Independent of the type of the distribution function (e.g. normal distribution, equal distribution)
- Stating the consistency on the basis of a fixed scale between "complete divergence" and "complete consistency"

These criteria are met by weighting the associated probability densities against each other; the associated densities are known from the specifications of the data sources. Standardization of the overlapping benchmark K_r, in relation to the individual values of the signals μ_i, is performed in order to describe the confidence within a fixed value scale, independently of the type of distribution function, the absolute uncertainty of the measurement signals and the number of available data sources. This is achieved by dividing the result of the same calculation by signals assumed to be overlapping in an ideal setting. The differences in $\mu_i = 0$ are set for this:

$$K_r = \frac{\int_{-\infty}^{+\infty} \prod_{i=1}^{i=n} p_i(x, \mu_i, \sigma_i) \cdot dx}{\int_{-\infty}^{+\infty} \prod_{i=1}^{i=n} p_i(x, 0, \sigma_i) \cdot dx} \qquad 0 \leq K_r \leq 1 \tag{10.1}$$

Where

K_r stands for the relative confidence benchmark / overlapping benchmark, no unit
n stands for the number of compared signals
p_i stands for the probability density function of the i-th signal
In the normal distribution shown here:
μ_i: Average of the probability distribution
σ_i: Standard deviation of the probability distribution

10.5.2.2.2 Dispersion Benchmark / Absolute Confidence

The goal of this part of the confidence benchmark is to describe the general uncertainty level of all available measurement signals on the basis of an absolute value, and hence to obtain the maximum achievable level of accuracy in the confidence test. The requirements for the consistency benchmark are derived from these criteria:

- Signals with high uncertainty make a minor contribution to improvement; signals with low uncertainty make a major contribution to improvement.
- The result states the best possible confidence level, independent of the actual overlapping of the probability densities.
- The benchmark shall be independent of the operating point and values of the signals.
- The benchmark shall be dependent on the number of available signals, as every data source that comes along improves the evaluation as a further option for checking.

These criteria can be met by examining the multivariant probability densities. In the case of independent, normally distributed dispersions, the variance in the dispersion of the fused signal can be formed by reciprocally adding individual variants (see Eq. (10.2)).

$$K_a = \left[\sum_{i=1}^{i=n} \sigma_i^{-2} \right]^{-\frac{1}{2}} \tag{10.2}$$

Where

K_a stands for the absolute confidence benchmark / dispersion benchmark, in the unit of the standard deviations
n stands for the number of compared measurement signals
σ_i stands for the standard deviation in the i-th signal

In order to take other probability density functions, and even dispersion dependencies, into consideration, the multivariant probability density must be formed in each case, and the dispersion benchmark determined in relation to a limit value. This results in a benchmark K_a that depends only on the available signals but not on the individual values of the signals and that evaluates the absolute accuracy of the integrity test.

10.5.2.2.3 Example: Consistency Benchmark and Dispersion Benchmark

Three exemplary probability densities are shown in Fig. 10.4. The expected value corresponds to the measured value in the relevant data source in the measurement epoch under examination; the dispersion around this measurement value corresponds to the specified uncertainty in the associated data source. The values are normally distributed, where:

$P_1 = N(0, 1)$ represents a source with average uncertainty
$P_2 = N(2, 0.5)$ represents a source with low uncertainty and deviation in the measurement signal from the other sources
$P_3 = N(0, 3)$ represents a source with high uncertainty and without offset to p_1

The results of the confidence calculations of the different distributions in accordance with the description in Sects. 10.5.2.2.1 and 10.5.2.2.2 are given in Table 10.2.

The consistency benchmark describes, within the standardized range between 0 and 1, the degree of consistency, where 0 stands for divergence and 1 for consistency.

Table 10.2 Values for confidence benchmarks

No.	Compared probability densities	Consistency benchmark K_r	Dispersion benchmark K_a
1	$p_1 \cdot p_3$	1.0000	0.9487
2	$p_1 \cdot p_2$	0.2019	0.4472
3	$p_2 \cdot p_3$	0.8056	0.4932
4	$p_1 \cdot p_2 \cdot p_3$	0.1757	0.4423

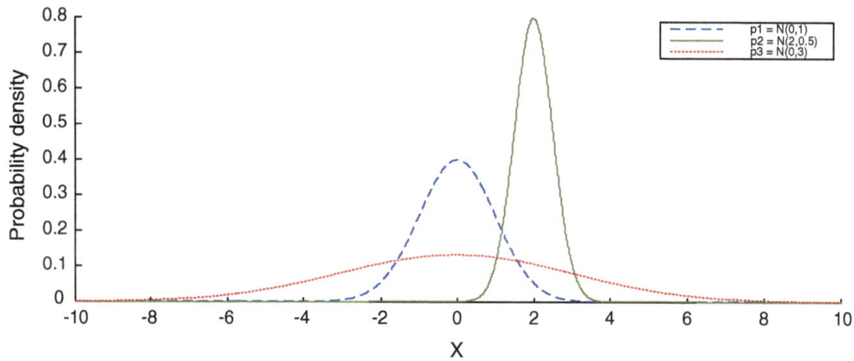

Fig. 10.4 Probability densities (Dziubek et al. 2012, p. 12)

By contrast, the smaller the dispersion benchmark value is, the more positively it is to be evaluated, as it has the unit of a standard deviation. Comparisons 1 to 4 show that the calculated values K_r and K_a behave in accordance with the specification. A user function thus receives an extended description of the binary integrity statement on the basis of both the actual degree of consistency of the current measurement values, as well as the the maximum overall accuracy level determined by the availability of data sources. The unambiguity and the underlying confidence of the integrity evaluation have thus been described for the user function. Although the desire to combine the two values into a "good/bad" statement seems obvious, this is problematic. This is also indicated by the difference in units ($[K_r] = 1$, $[K_a] =$ [Signal]). Only the statements in the two "corners" $K_r \approx 1$, K_a low and $K_r \approx 0$, K_a high, can be used to come to a clear conclusion regarding the quality (high or low).

10.5.2.2.4 Summary: Signal Confidence

The confidence benchmarks shown are elements in the expandable integrity and accuracy benchmark, which can be adapted to meet the requirements of the current architecture, and they evaluate the integrity statement with additional pieces of information. The initial binary evaluation concerning whether the compared measurement signals lie within plausible limits in accordance with the specification, receives additional information in the form of the consistency benchmark, which describes the relative deviation in the measurement signals from each other, and the dispersion benchmark, which contains an absolute statement regarding the accuracy level of the measurement signals available for comparison and evaluation. Simultaneous examination of the relative signal description dependent on the measurement value, and of the absolute signal description dependent on the availability of data sources, allows the confidence of the integrity statement to be described in the form of consistency between measurement signals, as well as in the form of the accuracy level that can be achieved with these measurement signals. They thus offer basic information for making a decision on the usability of the data, even in safety-critical systems. The modular principle of the confidence description allows additional benchmarks to be added without necessitating changes to the existing blocks.

If the results of the fusion filter—regarded as a black box—are also included in the comparison, this enables also checking the filter with its input data. As a result, possible instabilities in the filter that contradict its input data can be recognized. Such a system is presented in the following section.

10.5.2.3 Using Inherent Filter Variables to Evaluate the Integrity

A further possible element for the integrity evaluation is an algorithm integrated into the fusion filter itself. This concept is basically derived for utilization in the fusion filter presented in Sect. 10.4, but can be used for other filter types as well.

10.5.2.3.1 Approach

The approach is based on recursive filter types such as the error state space Kalman filter presented above. In general, these filters have been structured in such a way that a deviation is calculated by forming a difference between the expected value of a filter state (prediction) and the measured reference value (correction measurement), and converted into the relevant units by means of a system model if necessary. This difference is called the *innovation* $\vec{i}$. In general, this innovation is weighted with the uncertainty of the corresponding state and the uncertainty of the measurement before it is attached as a correction for the system state.

10.5.2.3.2 Derivation of the Evaluation Variables

As every measurement value incorporated into the filter also leads to an associated innovation, it is safe to assume that the innovations can be used as a benchmark for error detection as errors in the system states as well as in the measurement variables lead to deviations from each other, and hence to larger innovations. Usage of the *innovation* $\vec{i}$ as an evaluation variable is particularly suitable for the following reasons:

- The expected value of the innovation if there are no errors is $E\{i\} = 0$. This simplifies the formation of threshold values for the detection of errors.
- As the innovation constitutes a difference from the variables and measurement values inside the filter, errors in the filter and errors in the measurement values both lead to increased innovation values.
- The innovations have not yet been weighted by the filter's stochastic model. As a result, a consequent separation of filter and monitoring is achieved even though the error detection is incorporated into the filter.

- The monitored variable is completely consistent with the fusion filter as the filter's system model that converts the measurement variables into the units that match each other (analytical redundancy) is also used to calculate the evaluation variables.

As Kuusniemi (2005, p. 71ff), shows, a parameter can be calculated on the basis of the innovations by means of suitable weighing. This parameter takes both the uncertainty of the measurement values as well as the maximum false alarm rate required for the design into consideration. A comparison criterion is created by specifying a threshold value for the parameter. The binary statement "Error detected / no error detected" is made on the basis of this criterion.

10.5.2.3.3 Determining the Maximum Uncertainty

Specifying a threshold value for the error detection implicitly specifies the largest error that cannot yet be detected. If used in the data fusion in the filter, this error in the *measurement variable* leads to a deviation in the filter's fused *result*. According to Liu et al. (2011, p. 1785ff), this makes it possible to determine a confidence interval in a worst-case scenario within which the real result is very likely to be found.

This calculation is divided into three steps:

- Calculation of all the largest innovations currently not recognized as errors.
- Calculation of the effect on the result of the data fusion when such an error arises. The number of errors expected to arise at the same time must be borne in mind here.
- Searching for the maximum deviation based on the condition of the simultaneous number of errors.

The maximum deviation calculated in this way defines the confidence interval of the fusion result, and also uses filter-internal stochastic model variables here in order to realistically reproduce the actual effects of an error on the variable to be monitored.

10.5.2.3.4 Slowly Growing Errors

The method described is a suitable means for detecting abrupt errors. Slowly growing errors from the IMU, for example, can cause the solution to drift away as the Kalman filter generally adapts to the slowly growing errors.

One method of detecting this error type as well described by Bhatti and Ochieng (2009, p. 1785ff), is to check the change rate (smoothed by a pre-filter) in the weighted innovations against an error detection threshold.

10.5.2.3.5 Classification of the Integrity Evaluation

The integrity evaluation incorporated into the fusion filter provides further elements that complement the general integrity test approach. The result of the check as well as the other concepts presented provide a binary statement regarding the assumed absence of errors from the system, as well as a confidence interval within which the true value of the monitored variable is located with a low, known error probability. Both the measurement variables as well as the fusion filter itself are monitored. This approach thus meets the requirements for integrity evaluation given in Sect. 10.5.2.

Thanks to integration into the filter, a maximum degree of consistency in the redundancy checking and the system model of the filter is achieved. The disadvantage in this approach is the absence of total independence from the filter. However, this approach can be used with most filter types that work with innovations and stochastic descriptions.

10.5.3 Accuracy

10.5.3.1 Accuracy Statement

In the integrity evaluation using stochastic modelling, only limit values for evaluation of the system state can be calculated. By contrast, the goal of the accuracy statement is to calculate signal-specific parameters that describe their actual characteristics. Adaptive filters such as the error state space Kalman filter that change their characteristics over time are frequently used in a data fusion. Control devices that use the signals output by the filter, e.g. to control the driving dynamics, therefore need information about the current characteristics of the signals.

In general, the characteristics of the measurement signals output by a data source are known from the source's specification or have been calculated by means of measurements. In order to meet the requirement of a correct description of the processed signal, an approach in which a signal description is designed starting from the source is selected, and updated by all systems involved in the signal processing chain depending on their individual, internally known parameters. Every block has inputs and outputs for this set of description parameters, but is otherwise regarded as a black box. Modelling of the data processing by means of a general error propagation calculation allows both absolute, maximum errors as well as typical, average errors to be calculated.

The parameters used for this description are derived from typical information (Niebuhr and Lindner 2002, p. 13ff, p. 129ff) in sensor data sheets and standard sensor error models, as described in Wendel (2007, p. 194). The calculations for further processing in the signal chain are derived from basic operations in digital signal processing. The error classes named in Sect. 10.5.1 are used as the basis.

10.5.3.2 Parameters and Processing

10.5.3.2.1 Signal Description

The selected set of parameters for the description of a data signal is made up of the following:

- Noise: Stochastic, zero-mean error
- Offset/bias: Additive error
- Scale factor: Multiplicative error
- Non-linearity: Depends on the operating point, maximum error stated
- Offset and scale factor drift: Maximum specified change rate over time
- Bandwidth/cut-off frequency: Frequency range of the useful signal, and of disturbance suppression if necessary
- Delay time: Average group delay within the useful signal bandwidth
- Dispersion over time in the useful signal's frequency range

The unprocessed raw signals of the data sources form the starting point for the description of these parameters. These are modified during further processing of the signals. This processing is subdivided into four basic operations named below. A calculation based on the error class and the error propagation calculation have to be used.

- Error propagation in the event of ***addition/subtraction*** of the signal Modelling as an ideal superposition/addition element
- Error propagation in the event of ***amplification/attenuation*** of the signal Modelling as an ideal P-element/multiplication element
- Error propagation in the event of ***integration*** of the signal: Modelling as an ideal I-element/summation element
- Error propagation in the event of ***differentiation*** of the signal Modelling as an ideal D-element/differentiation element

The behavior of linear, time-invariant systems (Horn and Dourdoumas 2004, p. 18ff) can be described using these basic operations.

Particularly with regard to usage in time-discrete systems, and their filter operations based on unit delay elements described in the z-transformation (Horn and Dourdoumas 2004, p. 115ff), a fourth operation is introduced:

- Error propagation in the event of ***delaying*** of the signal Modelling as an ideal unit delay element (z^{-1})

By combining these operations with each other, any linear signal processing steps can be achieved, e.g. by means of state space modelling. Linearization also makes it possible to describe non-linear systems if the errors are small enough.

10.5.3.2.2 Summary of Accuracy

The accuracy statement in the form of a description similar to typical data sheets allows measurement signals to be characterized from their source, through the signal processing chain and right up to their sinks. Unlike the integrity description in which compliance with the limit values of the entire data collection is evaluated, the accuracy description addresses the actual characteristics of individual measurement signals. To this end, the accuracy evaluation uses a description from data sheets to calculate the signals' static specification on a one-off basis, and, if necessary, to calculate a dynamic specification on a continuous basis.

These description data can be evaluated in accordance with the user function's requirements. This means that adaptive control is possible or accuracy-dependent additional functions of driver assistance systems can be offered.

10.6 Conclusion, Outlook

On the basis of the requirements described in Sect. 10.3.1 to be met by a fusion of heavily heterogeneous signals, a general architecture was derived in the top-down procedure. This architecture takes the form of a centralized platform that allows the separation of the common direct linking of data sources and applications. This results in a standardized interface for both the sources and the applications. This interface considerably simplifies the usage, changes and additions in the system structure as well as the complexity of the application. Central processing also makes it easier to better detect erroneous measurement data by checking redundant data.

In addition, a fusion filter approach that corresponds to the structure of the architecture, and therefore meets the criteria for fusion of heavily heterogeneous signals, has been shown. This filter for fusing navigation data has a universal, 3D system structure. In addition to the data sources shown, this allows additional sources to be added, such as map matching, DGPS, WLAN/RFID positioning, radar or camera. Thanks to the concept of virtual sensors, which feed vehicle-specific assumptions into the filter in the same form as measurement data, using the filter structure in vehicle types with different driving dynamics, e.g. in motorbikes, can be achieved by simply replacing the virtual sensors. Thus, it could even be used in maritime and aerospace applications.

The requirements and criteria shown in Sect. 10.5.1 for describing the signal quality form the basis for creating a division of a signal quality description into integrity and accuracy by means of a top-down methodology, and the corresponding benchmarks are specified. The structure of the fusion architecture results in a decoupling that does not allow the applications any direct access to the data sources. This means that signal plausibilization at application level is no longer possible. In order to rectify this loophole and to use the advantages arising from the mutual testability of the data sources, the data quality description is added to the data fusion. Just like data fusion, the signal quality description is created from individual, adaptable and expandable components.

Both fusion and quality description can be adapted to different environment architectures thanks to the modular structure. The necessary characteristics of the fused data and quality description can be derived from the requirements for the most demanding application that uses the relevant data. This leads to a systematic approach to create a complete data processing architecture, and even to retrospectively integrate it into existing architectures with little outlay.

In addition, a quality evaluation integrated into the fusion filter is derived which also meets the existing requirements. It can be used in other, similar fusion filter types. However, additional components can be created, depending on the application, to expand the quality evaluation or to use it in completely different filter structures.

The accuracy description and error propagation calculation derived on the basis of linear basic signal processing operations allows an accuracy benchmark to be calculated for most applications. However, other, e.g. non-linear, models can be added to this description if necessary.

The combination of sensor data fusion and quality evaluation of the data thus allows encapsulation and replaceability of the architecture block "Sensor fusion/ virtual sensor" largely independently of hardware and software. It also complements and simplifies the detection of errors, and thus leads to possible simplification of existing user functions, and facilitates innovative functions that cannot yet be created with the current architecture.

References

Bar-Shalom, Y., Li, X., Kirubarajan, T.: Estimation with Applications to Tracking and Navigation, vol. 1. Wiley, New York (2001)

Bhatti, U.I., Ochieng, W.: Detecting multiple failures in GPS/INS integrated system: a novel architecture for integrity monitoring. J. Glob. Position. Syst. **8**(1), 26–42 (2009) (Centre for Transport Studies, Department of Civil and Environmental Engineering, Imperial College, London)(2009)

Bickford, R.L., Bickmore, T.W., Caluori, V.A.: Real-Time Sensor Validation for Autonomous Flight Control. The American Institute of Aeronautics and Astronautics, Reston (1997)

Brown, R.G.: A Baseline RAIM scheme and a note on the equivalence of three RAIM methods. In: Proceedings of the 1992 National Technical Meeting of the Institute of Navigation, San Diego, California (1992)

Dziubek, N., Winner, H., Becker, M., Leinen, S.: Fahrstreifengenaue Ortung von Kraftfahrzeugen durch Datenfusion und Fehlerkompensation von Standard-Seriensensoren. In: DGON-Symposium Positionierung und Navigation für Intelligente Verkehrssysteme, POSNAV ITS 2011, Darmstadt (2011)

Dziubek, N., Winner, H., Becker, M., Leinen, S.: Sensordatenfusion zur hochgenauen Ortung von Kraftfahrzeugen mit integrierter Genauigkeits- und Integritätsbewertung der Sensorsignale. In: 5. Tagung Fahrerassistenz, TÜV Süd - 5. Tagung Fahrerassistenz: Schwerpunkt Vernetzung, Munich (2012)

Feng, S., Ochieng, W.: Integrity of navigation systems for road transport. In Proceedings 14th World Congress of Intelligent Transportation Systems, Beijing (2007)

Goebel, K., Agogino, A.: Fuzzy sensor fusion for gas turbine power plants. In Proceedings of SPIE, Sensor Fusion: Architecture, Algorithms, and Applications III, vol. 3719, 7–9 April '99, Orlando, Florida (1999)

Horn, M., Dourdoumas, N.: Regelungstechnik. Pearson Verlag, Munich (2004)

Ibargüengoytia, P.H., Sucar, L.E., Vadera,S.: Real-time intelligent sensor validation. IEEE Trans. Power Syst. **16**(4), 770-775 (Atlanta, Georgia) (2001)

Kammeyer, K.D., Kroschel, K.: Digitale Signalverarbeitung. B. G. Teubner, Stuttgart (1998)

Kuusniemi, H.: User-Level Reliability and Quality Monitoring in Satellite-Based Personal Navigation. Tampere University of Technology, Finland, Institute of Digital and Computer Systems, Tampere (2005)

Leinen, S.: Parameterschätzung I / Parameter estimation I. Institute of Physical Geodesy, TU Darmstadt, Darmstadt (2010)

Le Marchand, O., Bonnifait, P., Ibañez-Guzmán, J., Bétaille, D.: Vehicle Localization Integrity Based on Trajectory Monitoring. Intelligent Robots and Systems, St. Louis, Missouri (2009)

Liu, J., Tang, T., Gai, B., Wang, J.C.: Integrity Assurance of GNSS-Based Train Integrated Positioning System. Beijing Jiaotong University, Beijing, Science China Press and Springer-Verlag, Berlin, Heidelberg, State Key Laboratory of Rail Traffic Control and Safety (2011)

Mansfeld, W.: Satellitenortung und Navigation, 2nd edn. Vieweg Verlag, Wiesbaden (2004)

IS-GPS-200: Interface Specification Revision D, Space and Missile Systems Center (SMC), Navstar GPS Joint Program Office (SMC/GP). El Segundo, California (2006)

Niebuhr, J., Lindner, G.: Physikalische Messtechnik mit Sensoren, vol. 5. Oldenbourg Industrieverlag, Munich (2002)

Pourret, O., Naim, P., Marcot, B.: Bayesian Networks: A Practical Guide to Applications. Wiley, West Sussex (2008)

Soika, M.: A sensor failure detection framework for autonomous mobile robots. In: Proceedings of the 1997 IEEE/RSJ International Conference on Intelligent Robots and Systems, Grenoble (1997)

Toledo-Moredo, R., Zamora-Izquierdo, M.A., Úbeda-Miñarro, B.: High-Integrity IMM-EKF-Based Road Vehicle Navigation with Low-Cost GPS / SBAS / INS. IEEE Trans. Intell. Transp. Syst. **8**(3), 491–511 (Università di Parma, Parma) (2007)

Wendel, J.: Integrierte Navigationssysteme. Sensordatenfusion, GPS und Inertiale Navigation, Oldenbourg Wissenschaftsverlag, Munich (2007)

Young, R.S.Y., McGraw, G.A.: Fault detection and exclusion using normalized solution separation methods. In Proceedings of the 15th International Technical Meeting of the Satellite Division of the Institute of Navigation, Portland, Oregon (2002)

Zogg, J.-M. (2009): GPS und GNSS: Grundlagen der Navigation und Ortung mit Satelliten (updated Oct. 2011), μBlox AG, Thalwil

Part V
Functional Testing

Chapter 11
Testing of Reconfigurable Systems: A Cognitive-Oriented Approach

Asem Eltaher

11.1 Introduction

Generally, a reconfigurable system is a component-based system that consists of several hardware and software components that are independently developed either on-site, or by third parties (Denaro et al. 2003). As a result, reconfigurable systems involve the ability to replace one or more component(s) of the Device-Under-Test (DUT), which permits new possible configurations that make the test process an expensive burden. Indeed, most of the existing test techniques are foiled by the assumption that the internal structure of the DUT is known with at least partial access to the source code (Bezerra et al. 2001).

In the reality, the assumption of complete knowledge about a reconfigurable system is not generally true since several components may be supplied by a third party. Therefore, testing of reconfigurable DUTs represents new challenges that cannot be adequately manipulated by traditional testing techniques. In this context, facing the fact that "it is impossible to fully test a product" (Kaner 1997); one central issue is how to define a reasonable end to testing processes. Traditional solutions, e.g. Dalal and Mallows (1992), Gemoets et al. (1994) and Levendel (1990), end a testing cycle when the odds of detecting additional faults are below a certain threshold.

Related theories tend to consider bugs as discrete objects, statistically allocated over the entire software space with a known distribution (Levendel 1990). In an effort to achieve further improvements, a statistical optimization approach is developed in Dalal and Mallows (1992) to extend this work to unknown distributions and enrich it by some graphical aids. Later on, research is done in Gemoets et al. (1994) to consider the choice of fuzzy reliability models as an optimization problem and solve it.

Indeed, on one side, approaches in Dalal and Mallows (1992), Gemoets et al. (1994) and Levendel (1990) offer reasonable theoretical explanations for the test

A. Eltaher (✉)
Institute of Control Engineering,Technische Universität Braunschweig,
Hans-Sommer-Str. 66, D-38106 Braunschweig, Germany
e-mail: eltaher@ifr.ing.tu-bs.de

M. Maurer and H. Winner (eds.), *Automotive Systems Engineering*,
DOI: 10.1007/978-3-642-36455-6_11,

results. But, on the other side, they inherit auxiliary problems to capture some qualitative aspects; e.g. how to define a bug, how to report bugs, the severity of an error, etc.

Moreover, classical automatic testing methods discard the ability of skilled human testers to enrich test processes with intuition-based strategies. In response to this limitation, the last decade witnessed an emerging interest in studying human intelligence to mimic it during automatic testing. For example, the contribution in Gras et al. (2006) highlights an approach to interview skilled human testers to get an insight into their experiences. These experiences formulate the training sets for a learning test system, based on Bayesian Networks (BNs), which directs the test session to the most likely defect areas in the DUT.

In spite of promising preliminary results, this approach entails an expensive burden on human testers since "it requires normally the analysis of a large number of cases, covering almost every possible combination of input variables" (Gras et al. 2006). Furthermore, the theory of fuzzy logic shows that humans describe their experiences in imprecise and vague language that can hardly be formally described (Zadeh 1973).

In response to the above challenges, this chapter outlines a novel approach to develop learning test systems, which observe skilled human testers during various test sessions. Consequently, observed test-sessions are modeled, optimized, and generalized to be further applied in similar test situations.

The core assumption of the above stated approach is the ability of skilled human testers to design and adapt a significant range of test inputs under which a failure may arise. Then, they proceed further till they reach a decision that additional testing will not significantly change the test results. This process is defined in Bach (1998) as "good enough testing". The realization of the proposed approach is outlined in the following sections. Section 11.2 illustrates the structure of the learning environment. Next, Sect. 11.3 describes the evaluation metrics used to benchmark the performance of the test system. Section 11.4 shows the migration process from learning to testing, whereas, Sect. 11.5 addresses achieved results. Finally, Sect. 11.6 summarizes this contribution and future work.

11.2 Learning Test Systems

11.2.1 Human-Machine-Interaction (HMI): Observation

Indeed, the core conception is the assumption that HMI is based on a perception-action concept (see Fig. 11.1). In the *perception* phase, the human tester perceives information about the DUT; also defined in Bach (1998), Gras et al. (2006) as situation awareness. Then, the *action* phase takes place, in which human testers decide what the next best action(s) is (are).

Briefly, the test system interprets a test session as a series of perception and action phases. And the next step is to model the observed HMI, as a prerequisite to store the data, which is illustrated in Sect. 11.2.2.

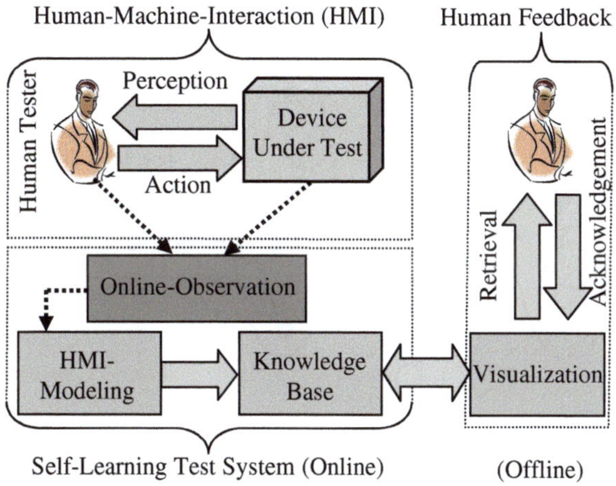

Fig. 11.1 Learning by observing human testers

11.2.2 Human-Machine-Interaction (HMI): Modeling

The choice of the modeling technique is based on the assumption that HMI can be *observed* and *modeled* as an event-driven process. Situation-Operator-Model (SOM) in Söffker (2001) is the adopted modeling technique. SOM models the changes of the considered part of an external environment as a sequence of effects. These effects are described by the items *scenes* and *actions*. A real world *scene* is modeled by a *situation,* whereas, an *action* is modeled by an *operator*.

The item *situation* (S) models the observed states of a DUT which consists of a set of characteristics (C) and relations (R). The characteristics are a set of representative elements such that each characteristic describes a definite part of the DUT. The introduced item characteristic, e.g. (C_k), offers the opportunity to define a time-dependent parameter (P_k) that describes the current observed state. A relation (r_i) describes the inner connection(s) between different characteristics of the same situation, if they exist.

The item *operator* (O) models the action(s), invoked by human testers which shift(s) the DUT from an initial situation (S_I) to a next/final one (S_F). Figure 11.2 shows a situation, changed by an operator that leads to another situation which is denoted in Söffker (2001) as an *experience*. Next, the current final situation is defined as the initial situation for the next experience and so on. Hereby, SOM offers a flexible framework to model a session of HMI as a sequence of test cases modeled by experiences.

A detailed description about the usage of SOM to model HMI during a test session is given in Eltaher et al. (2008). And, in the context of this contribution, SOM-terms are further developed as shown below:

Definition 11.1 Situation-Path (P): is the sequence of operators that lead to an arbitrary target situation (S_X) with respect to a predefined initial situation (S_I).

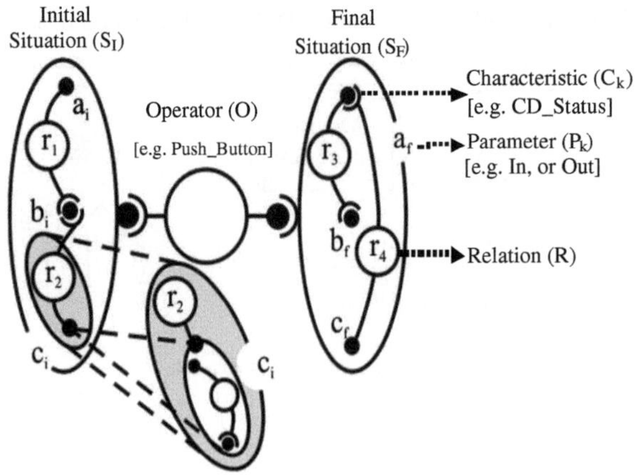

Fig. 11.2 Structure of Situation-Operator-Model (SOM) (Söffker 2001)

Definition 11.2 Two paths are equal, if they involve the exact type and sequence of operators to reach a certain situation. Obviously, considering a complex DUT, different paths can lead to the same situation.

For example, referring to Fig. 11.3, the following facts can be concluded:

- S_3 is reached via two different paths (P_1, P_2).
- P_1 involves the sequence $\{O_1, \{O_2, O_3\}, O_4\}$.
- P_2 involves the sequence $\{O_3, O_4\}$.

Finally, a test system is of no use, it is not able to store acquire experiences. Hence, a knowledge base module is implemented, as it is shown in Sect. 11.2.3.

11.2.3 Knowledge Base and Human Feedback

The objective of this module is to enable human testers to: (a) acknowledge the consistency of the stored experiences and (b) alter the stored data. Case (a) is triggered, if the test system announced the existence of inconsistent experiences in its knowledge base. This would be the case, if it observed the same test case with two different results, which may occur due to a human error.

On the other side, altering the knowledge base is demanded, if the DUT delivered wrong reaction(s) during test sessions. Driven by cases (a) and (b), the test system has been enriched with a 2D visualization module to facilitate the feedback process.

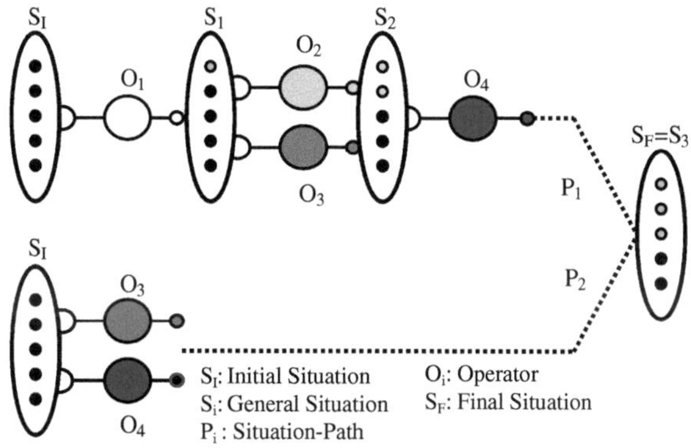

Fig. 11.3 Reaching the same situation via different paths

11.2.4 Cooperative Learning

Learning from a homogeneous knowledge source, i.e. a sole human tester, may suffer from the cognitive biases that have been found in human testers (Berndt et al. 2003). This can result in an under/over-exploration of some test segments, which leads to a poor metric of test coverage.

Given this premise, observing several human testers by the test system is accomplished to guard the learning process against over-fitting, or under-fitting. Besides, this leads to maximize individual benefits of each human tester. A detailed description of this module is given in a previous contribution (Eltaher et al. 2009).

11.3 Evaluation Metrics

11.3.1 Idea

Practically, it is interesting to evaluate the test strategies of human testers based on a defined set of metrics. This has to offer a double benefit for the proposed approach. First, it helps to compare the performance of human testers against each other. Second, it provides the means to benchmark the efficiency of the test system against human testers. This leads to a consistent evaluating of the extent to which an improvement has been achieved. In this chapter, the evaluation vector includes three quantitative metrics:

- Testing time,
- probability of success, and
- test coverage.

11.3.2 Evaluation Metrics: Testing Time (T)

In the real world, time constraints hinder the possibility to exercise all possible test segments. As a natural result, a certain time limit for the testing process has to be defined. Therefore, to model the reality, a time limit for each test session is given, which is set proportionally to the complexity of the DUT.

11.3.3 Evaluation Metrics: Probability of Success (P_E)

Certainly, fault coverage is a central aspect to estimate the efficiency of a test session (Wang et al. 2010). To incorporate this metric, an Error-Insertion-Module (EIM) is implemented to generate two diverse arts of errors, *immediate* and *latent* errors. An immediate error is the one that is always detectable, which is independent of the states/transitions coverage. Whereas, a latent error cannot be detected unless some other state(s)/transition(s) is (are) triggered.

Indeed, fault coverage cannot be calculated based on a sole test-session. This does not lead to a commitment that the corresponding human tester posses a test strategy with a consistent fault revealing ability. Therefore, each human tester executes Q test-sessions, in which the EIM modifies the type and location of the errors for each new session. Accordingly, fault coverage is experimentally calculated, which is denoted in this work as the probability of success.

Definition 11.3 Probability of success (P_E) spells the likelihood of a complete error detection by a human tester in the given time for a testing session(T).

Table 11.1 illustrates the algorithm, used to calculate (P_E) via estimating first the individual detection probability of the errors $E_1 \ldots E_N$ using an error detection matrix (e):

$$e = \begin{pmatrix} 1 & 0 & \ldots & 1 \\ 0 & 1 & \ldots & 0 \\ \ldots & \ldots & \ldots & \ldots \\ 1 & 0 & \cdots & 1 \end{pmatrix} \text{ according to Table 11.1}$$

Specifically, P_{Ei} is the number of times an error (E_i) has been detected divided by the total number of its appearance. Next, (P_E) is generally calculated as:

$$P_E = \sum_{i=1}^{N} \alpha_i \cdot P_{Ei} \tag{11.1}$$

Such that $\sum_{i=1}^{i=N} \alpha_i = 1$ and:

Table 11.1 Calculating the probability of success (P_E)

Errors/test-session (TS)	TS_1	TS_2	...	TS_Q	P_{Ei}
Error (E_1)	**1**	0	...	**1**	P_{E1}
Error (E_2)	0	**1**	...	0	P_{E2}
...	...	...	...	...	...
Error (E_N)	**1**	0	...	**1**	P_{EN}

P_{Ei} Detection probability of the error E_i, N Number of errors, Q Test-Sessions Number, 1 Error is detected, 0 Error is NOT detected

Table 11.2 Probability of success for human tester-A

Errors/test-session	TS_1	TS_2	TS_3	TS_4	P_{Ei}
Error (E_1)	**1**	0	0	**1**	0.5
Error (E_2)	**1**	**1**	**1**	0	0.75

- N : Number of generated errors
- α_i : Arbitrary weighting factor

Assuming that all errors are equally significant, then:

$$\alpha_1 = \alpha_2 = \alpha_N = \frac{1}{N} \tag{11.2}$$

Table 11.2 shows an example to calculate the probability of success for tester-A (P_{E_A}), given the following parameters:

- $Q = 4, N = 2, \alpha_1 = \alpha_2 = 0.5$, which leads to:
- $P_{E_A} = 0.5 \times 0.5 + 0.5 \times 0.75 = 0.625\ (Rounded : 63\,\%)$

Later on, performance of the test system against several human testers, e.g. (A) and (B), has to be compared. To this end, (e_{AB}) is the combined error detection matrix for testers (A) and (B) that is defined as:

$$e_{AB} = (e_A \cup e_B) - (e_A \cap e_B) \tag{11.3}$$

Next, ($P_{E_{AB}}$) is the combined probability of success for testers (A) and (B) that is calculated via applying Eq. (11.1) to the output matrix from Eq. (11.3).

11.3.4 Evaluation Metrics: Test Coverage (C)

Practically, rules of thumb are often used to define a complete test domain and its subsets (Ngamsaowaros and Sophatsathit 2007). To this goal, it is needed to address the following terms: complete coverage, individual coverage, and combined coverage.

Complete Coverage (CC)—It is the sum of all unique paths to reach all observed situations, starting from the predefined initial situation (S_I). For example, Fig. 11.4

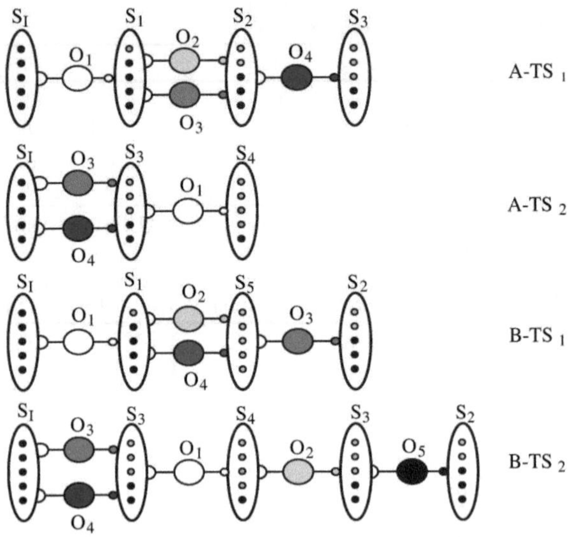

A-TS j: Tester-A scenario during the j th test-session
B-TS j: Tester-B scenario during the j th test session
$S_{1,2,3,4,5}$: Situations that model real world scenes
S_I: Initial situation
$O_{1,2,3,4,5}$: Operators that model testers' actions

Fig. 11.4 Test-scenarios from human testers (A) and (B)

Table 11.3 Estimating the complete coverage (CC)

Situation (S_j)	A-TS_1	A-TS_2	B-TS_1	A-TS_2	NOP	CC
Situation (S_1)	**1**	0	**1**	0	**1**	**9**
Situation (S_2)	**1**	0	**1**	**1**	**3**	
Situation (S_3)	**1**	**1**	0	**2**	**3**	
Situation (S_4)	0	**1**	0	**1**	**1**	
Situation (S_5)	0	0	**1**	0	**1**	

NOP Number of Unique Paths (See Definition 11.1), *CC* Complete Coverage, 1/2 Number of times situation S_i appeared in the test-session, 0 Situation S_i did NOT appear in the corresponding test-session

shows the test-scenarios by testers (A) and (B). Accordingly, Table 11.3 is derived, which leads to a CC of (11.9) paths.

Individual Coverage (C)—It is the *average* test coverage achieved by a human tester in the given time for a testing session (T). This can be formally described like:

Table 11.4 Individual coverage of testers (A) and (B)

Human-tester	Test-session	NOP	CC	C-TS (%)	C (%)
Tester-A	A-TS_1	**3**	**9**	$C_{A1} = 33$	$\mathbf{C_A = 28}$
	A-TS_2	**2**		$C_{A2} = 22$	
Tester-B	B-TS_1	**3**		$C_{B1} = 33$	$\mathbf{C_B = 39}$
	B-TS_2	**4**		$C_{B2} = 44$	

C-TS: Coverage achieved during a certain test-session (TS), *C* Individual average-coverage of the human tester

$$C_A = \frac{1}{Q} \sum_{i=1}^{i=Q} C_{A_i} \tag{11.4}$$

Such that:

- C_A : Average coverage achieved by tester-A
- Q : Number of performed test-sessions
- C_{A_i} : Coverage of tester-A in the ith test-session

Practically, C_{A_i} is the number of unique situation-paths, covered by tester (A) in the ith test-session. Accordingly, Table 11.4 is deduced via applying Eq. (11.4) to Table 11.3.

Combined Coverage—It is the *average* of the total test coverage, achieved by two human testers in the given time for both human testers ($2T$). This can be formally described like:

$$C_{AB} = \frac{1}{Q} \sum_{i=1}^{Q} C_{AB_i} \tag{11.5}$$

Such that:

- C_{AB} : Combined coverage of testers (A) and (B)
- Q : Number of test-sessions
- C_{AB_i} : Combined coverage for ith test-session

C_{AB_i} is calculated as follows:

$$C_{AB_i} = (C_{A_i} + C_{B_i}) - (C_{A_i \cap B_i}) \tag{11.6}$$

The term $(C_{A_i \cap B_i})$ spells reaching the same situation using the same path by testers (A) and (B) in the ith test session. For example, in Fig. 11.4, S_1 is reached via the same path in A-TS_1 and B-TS_1. Accordingly, Table 11.5 shows the combined coverage of testers (A) and (B).

Table 11.5 Combined coverage of testers (A) and (B)

Test-session	$C_{Ai} \Updownarrow C_{Bi}$ (%)	$C_{Ai} \cap C_{Bi}$ (%)	C_{ABi} (%)	C_{AB} (%)
A-TS_1	$C_{A1} = 33$	11	$\mathbf{C_{AB1} = 55}$	$\mathbf{C_{AB} = 50}$
B-TS_1	$C_{B1} = 33$			
A-TS_2	$C_{A2} = 22$	22	$\mathbf{C_{AB2} = 44}$	
B-TS_2	$C_{B2} = 44$			

11.3.5 Evaluation Metrics: Fitness Function

Generally, lacking quantitative metrics makes the process of comparing the performance of human testers against each other quite challenging. As a result, it is needed to define a fitness function (f) for each human tester as follows:

Definition 11.4 Fitness Function (f): It defines the reliability of a human tester to orchestrate a test session with a significant power of fault revealing and feasible test coverage, in the given testing time (T).

In this work, (f) is calculated like:

$$f_A = \alpha_1 \cdot P_{E_A} + \alpha_2 \cdot C_A \tag{11.7}$$

Such that:

- f_A : Fitness function of tester-A
- P_{E_A} : Probability of success of tester-A (Eq. 11.1)
- C_A : Test coverage achieved by tester-A (Eq. 11.4)
- α_i : Arbitrary weighting factor

Obviously, the fitness function does not consider the testing time (T) as it is equal for all testers. Here, it is assumed that human testers fully consume the allowed time. Next, it is interesting to investigate the performance of the test system against human testers (A-B) in terms of the cost of both sides. Hence, it is needed to define a cost function (δ) for human testers (A and B) as shown below:

$$\delta_{AB} = \alpha_1 \cdot \frac{T_{AB}}{2T} + \alpha_2 \cdot (1 - C_{AB}) + \alpha_3 \cdot (1 - P_{E_{AB}}) \tag{11.8}$$

Such that:

- δ_{AB} : Cost function of the testers (A) and (B)
- T_{AB} : Testing time by (A) and (B)
- T : Testing time per session for each tester
- C_{AB} : Test coverage by (A) and (B). (see Eq. 11.5)
- $P_{E_{AB}}$: P_E for testers (A) and (B). (see Eq. 11.3)

- α_i : Arbitrary weighting factor

Typically, T_{AB} is equal to $2T$ under the previously mentioned assumption of a 100 % testing time consumption.

Though reaching a framework to record and evaluate test scenarios, executed by skilled human testers, the ultimate goal is not reached yet. The idea here is how to use the test scenarios to generate an optimal test oracle for the DUT, which is described in Sect. 11.4.

11.4 From Learning to Testing

11.4.1 Idea

In fact, the absence of theoretical bases to reduce the test domain will make the choice of the test scenarios with little, or no justification. To overcome this limitation, the test system has been enriched with a rule-based-reasoning (RBR) paradigm to manage the test execution process to the most likely defect areas in the DUT. The prior knowledge for the RBR-module invokes three significant rules. The first rule (R_1) aims to assign a quality factor (q) to the individual scenarios by testers (A) and (B) in Fig. 11.4.

The second rule (R_2) is designed to combine the best scenario from tester (A), i.e. the scenario with the highest (q), with the best one from tester (B) to develop new self-generated test scenarios.

Next, the third rule (R_3) picks the best-possible scenario from the self-generated scenarios, resulted by (R_2). To this goal, (R_3) assigns a risk factor (η) to each of the self-generated test scenarios. In Sect. 11.4.2, the application of the defined rules (R_1, R_2, and R_3) is illustrated with an example to the test scenarios, shown in Fig. 11.4.

11.4.2 Example

Briefly, (R_1) evaluates the individual scenarios from testers (A) and (B) based on the following formula:

$$q = \alpha_1 \cdot n + \alpha_2 \cdot c \tag{11.9}$$

Such that:

- q: Quality factor of a test scenario
- n: Number of found errors
- c: Test coverage achieved by a test scenario
- α_i : Arbitrary weighting factor

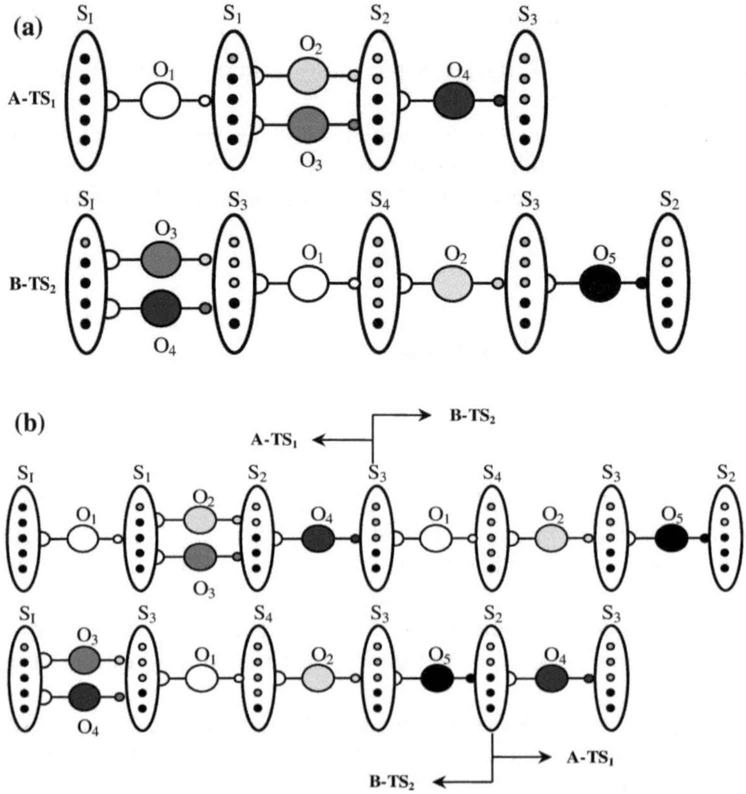

Fig. 11.5 **a** Best test scenarios from tester (A) and (B). **b** Structure of new self-generated test scenarios

Obviously, the time factor is not considered because all scenarios are executed by human testers in the same given testing time (T). So, referring to Fig. 11.4, the output of (R_1) is both scenarios (A-TS_1) and (B-TS_2) as shown in Fig. 11.5a.

Next, (R_2) plays the role to combine both test scenarios in Fig. 11.5a. To this end, the notion of task relatedness has to be defined, which is the common test case(s) between the two scenarios, i.e. A-$TS_1 \cap$ B-$TS_2 \neq \phi$.

Practically, the existence of common test case(s) among diverse test scenarios is not a naive assumption. It has been practically shown that test scenarios performed by skilled human testers—regardless their test strategy—share some test cases that aim to stimulate the basic functions of the DUT. Figure 11.5b shows the output of the second reasoning rule (R_2).

Finally, (R_3) assigns a risk factor (η) to each test scenario of the new ones in Fig. 11.5b as follows:

$$\eta = \alpha_1 \cdot t + \alpha_2 \cdot (1 - p) \tag{11.10}$$

Such that:

- η : Risk factor
- t: Execution time
- p: Fault detection probability
- α_i : Arbitrary weighting factor

For evaluating self-generated test scenarios, the execution time factor (t) is considered since the scenarios are not equal in length. Indeed, each test case has been observed by the test system at least once in the learning phase (Sect. 11.2). Hence, it is practically possible for the test system to roughly calculate the time, needed to execute each test case and, accordingly, each new self-generated test scenario.

In the reality, it has been shown that reaching the same situation via new situation-paths is an adequate approach to reveal more latent errors (Sect. 11.3.3). This is especially true since new test paths exercise some test segments in the DUT, which have not been exercised before. Motivated by this fact, fault detection probability (p) in Eq. (11.10) is the number of situations, reached by new paths divided by the total number of observed situations.

For example, considering the test scenario [A-TS_1-B-TS_2] in Fig. 11.5b, situations ($S_{4,3,2}$) are reached by new paths; while the total number of situations is five (see Fig. 11.4). Accordingly, this scenario has a value of $p = 60\,\%$. Likewise, considering the scenario [B-TS_2-A-TS_1], situation (S_3) is reached via a new path. Hence, it has been assigned the value of $p = 20\,\%$. Finally, each test scenarios is assigned a risk factor (η) based on Eq. (11.10).

Consequently, the test generator module plays the final role to execute the following test scenarios:

- A-TS_1, or B-TS_2, (the one with the higher q) and
- the self-generated scenario with the lowest (η).

In this context, to transit from one test scenario to another, it is assumed that the DUT-specifications provide a reset function from each situation to the initial one (S_I).

11.4.3 Test System versus Human Testers

As mentioned before, the ultimate goal is to compare the test system against the human testers (A) and (B). To this end, similar to Eq. (11.8), a cost function of the test system is defined as:

$$\delta_{sys} = \alpha_1 \cdot \frac{T_{sys}}{2T} + \alpha_2 \cdot (1 - C_{sys}) + \alpha_3 \cdot (1 - P_{E_{sys}}) \tag{11.11}$$

Such that:

- δ_{sys} : Cost function of the test system
- T_{sys} : Testing time consumed by the test system

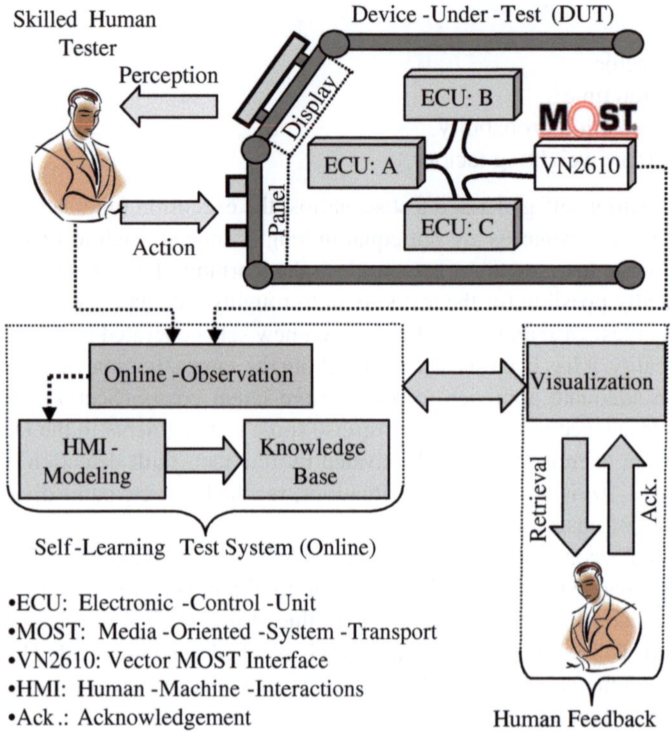

Fig. 11.6 Structure of the realized test system

- T: Testing time per session for each tester
- C_{sys} : Test coverage by the test system
- $P_{E_{sys}}$: Probability of success for the test system
- α_i : Arbitrary weighting factor

Reaching this stage, complete coverage (CC) has to be updated since the test system developed new paths for some situations. Accordingly, Tables 11.4 and 11.5 have to be updated.

Then, the coverage of the test system is calculated exactly like the coverage of a human tester.

Next, to be consistent, the test system executes the same number of test sessions (Q). And, for each new session, the Error-Insertion-Module (EIM) changes the type and location of the errors. This way, the probability of success for the test system ($P_{E_{sys}}$) is also experimentally calculated. Aside from the example in Fig. 11.4 that is used for clarification purposes, Sect. 11.5 shows the results in the real world.

Table 11.6 Cost function of human testers versus the test system

	Results	f	O(H)	O(SYS)
A	$PE_A = 0.72$ $C_A = 0.13$	$f_A = 0.43$	$PE_{AB} = 0.8$ $C_{AB} = 0.16$	$PE_{sys} = 0.92$ $C_{sys} = 0.36$
B	$PE_B = 0.72$ $C_B = 0.11$	$f_B = 0.42$	$T_{AB}/2T = 1$ $\delta_{AB} = 0.760$	$T_{sys}/2T = 1.1$ $\delta_{sys} = 0.730$
C	$PE_C = 0.68$ $C_C = 0.11$	$f_C = 0.40$	$PE_{CD} = 0.72$ $C_{CD} = 0.13$	$PE_{sys} = 0.84$ $C_{sys} = 0.32$
D	$PE_D = 0.6$ $C_D = 0.12$	$f_D = 0.36$	$T_{CD}/2T = 1$ $\delta_{CD} = 0.788$	$T_{sys}/2T = 1.17$ $\delta_{sys} = 0.795$
E	$PE_E = 0.6$ $C_E = 0.11$	$f_E = 0.36$	$PE_{EF} = 0.68$ $C_{EF} = 0.15$	$PE_{sys} = 0.8$ $C_{sys} = 0.29$
F	$PE_F = 0.56$ $C_F = 0.09$	$f_F = 0.33$	$T_{EF}/2T = 1$ $\delta_{EF} = 0.793$	$T_{sys}/2T = 1.23$ $\delta_{sys} = 0.843$
G	$PE_G = 0.52$ $C_G = 0.11$	$f_G = 0.32$	$PE_{GH} = 0.64$ $C_{GH} = 0.16$	$PE_{sys} = 0.76$ $C_{sys} = 0.28$
H	$PE_H = 0.48$ $C_H = 0.12$	$f_H = 0.30$	$T_{GH}/2T = 1$ $\delta_{GH} = 0.800$	$T_{sys}/2T = 1.33$ $\delta_{sys} = 0.905$
I	$PE_I = 0.4$ $C_I = 0.1$	$f_I = 0.25$	$PE_{IK} = 0.52$ $C_{IK} = 0.13$	$PE_{sys} = 0.64$ $C_{sys} = 0.21$
K	$PE_K = 0.32$ $C_K = 0.08$	$f_K = 0.20$	$T_{IK}/2T = 1$ $\delta_{IK} = 0.838$	$T_{sys}/2T = 1.37$ $\delta_{sys} = 0.973$

11.5 Realization and Results

Briefly, the realization of the online-observation module as well as Situation-Operator-Model (SOM) is done using *Microsoft C++*. Practically, observing HMI is accomplished by recording the communication data, sent over the Media-Oriented-System-Transport (MOST) data bus. To this goal, a MOST-based interface, named VN2610 in Fig. 11.6, is attached to the DUT.

The knowledge base is realized in *Python* and a free object-oriented database (*DyBASE)*. Whereas, the visualization module is realized using (*VPython*). Finally, an infotainment system is used as an experimental DUT. A more detailed description about the realization is given in Eltaher et al. (2008).

For demonstration purposes, ten human testers—from different ages with various experiences-participated in the experiments. Besides, the following parameters are selected:

- $\alpha_1 = \alpha_2 = 0.5$ for Eqs. (11.7), (11.9), and (11.10)
- $T = 5(\text{min})$, $Q = 5$, and $N = 5$ errors

For Eqs. (11.7), (11.9), and (11.10), the corresponding parameters are assumed to be equal in their importance. Accordingly, the weighting factors are set to be equal.

Table 11.6 shows the individual fitness functions of the human testers (A-K). Besides, the cost function of each two human testers O(H) is calculated versus the test system O(SYS) according to Eqs. (11.8) and (11.11) respectively for the following weighting factors: $\alpha_1 = 0.5$ and $\alpha_2 = \alpha_3 = 0.25$.

For Eqs. (11.8) and (11.11), the weighting factors are set so that the investment done by the system (time) is weighted equally to its benefits (errors found and test coverage). Hence, the timing parameter posses a weighting factor of 0.5. Likely, it is assumed too that both benefits (errors found and test coverage) are equally important. Accordingly, their total share of weighting (0.5) is equally divided between them. This would lead to an individual weighting factor of 0.25 for each factor.

Accordingly, results show that the test system performs dependent on the quality of the testers. Specifically, for human testers with relatively higher fitness function, the cost of the test system is less than, or almost equal to the cost of human testers and vice versa.

For example, in the given testing time (T), testers (A-B) possess a higher ability than testers (I-K) to orchestrate test scenarios with a significant power of fault revealing and good test coverage. According to Table 11.6, the results of tester (A) are the following:

- Probability of Success (P_E) is 72 % calculated according to Eq. (11.1).
- Test Coverage (C_A) is 13 % calculated according to Eq. (11.4).
- As a result, the overall fitness function (f_A) is 0.43 according to Eq. (11.7).

For each tester of the involved 10 testers, the individual fitness function is calculated as it is mentioned above. Then, the 10 testers are divided into 5 groups in which group 1 includes the best 2 testers and group 5 includes the worst ones. Finally, the test system is evaluated against each group. Considering group 1 (Testers A and B) as an example:

- Combined Probability of Success ($P_{E\text{-}AB}$) is 80 % calculated according to Eq. (11.3).
- Combined Test Coverage (C_{AB}) is 16 % calculated according to Eq. (11.5).
- As a result, the overall cost function δ_{AB} is 0.76 according to Eq. (11.8).

In the testing phase, the test system further optimizes the scenarios from (A) and (B) by choosing the best possible combination of the two scenarios that leads to reaching several situations using new paths. Consequently,

- more new latent errors are found ($P_{E_{sys}} = 92\,\%$),
- the test coverage has been significantly enhanced ($C_{sys} = 36\,\%$), and
- the corresponding invested testing time is 10 % more than testers A and B.

Therefore, though the extra time, invested by the test system to execute its self-generated test scenario, the cost function is still lower than the cost function of the human testers, i.e. the extra time is a worthwhile investment.

On the other side, combining relatively less efficient scenarios from testers (I-K), the extra time invested by the test system has not been paid off. This is especially true since only few more latent errors are found with relatively poor test coverage. In fact, the dependency between the test system's performance and the professionalism of human testers has 2 sides:

1. Using the test time efficiently.
2. Executing test scenarios that have a significant power of fault revealing.

In an effort to describe the timing aspect, the following case illustrates a real-world scenario. Indeed, based on the standard specifications of testing a compact-disc (CD) player, it is recommended to wait 3–7 s after switching from one track to another to make sure that the track is successfully changed. Skilled human testers like A and B follow such rules strictly during the training session. In this context, waiting more than 7 s to further proceed with the next test case is just time wasting.

Accordingly, the test system learns this testing behavior during the training session and applies this testing rule for any future test scenarios. This way, the test system uses its test time efficiently that leads to a cost-effective test strategy.

On the other side, relatively non-skilled human testers like I and K waited over 20 s after switching the track of a CD player. Indeed, the time frame of 20 s can be divided into 2 time frames:

- The first 7 s: Required and necessary waiting time.
- The last 13 s: Non-necessary waiting time.

Accordingly, the test system learns a sub-optimal testing behavior and applies this rule in the testing session. As a result, the extra time, invested by the test system is not a worthwhile investment. This clarifies why the test system needed only 10 % more time than the given standard testing time in case of testers A and B. Whereas, the test system needed 37 % more time than the given testing time in case of testers I and K.

Besides, unlike testers I and K, testers A and B have the ability to orchestrate test scenarios that enjoy a significant power of fault revealing. In this context, when the test system combines the corresponding test scenarios, it results in a very efficient test session that enhances the test coverage and increases the probability of success.

11.6 Conclusion and Discussion

According to the above results, for managing a test process, a crucial factor remains the test personnel. Beyond the availability of advanced tools and test methods, personnels' skills, commitment, and lessons learned play a significant role to test efficiently.

On the other hand, if human testers are not qualified, they are in *no* position to orchestrate efficient test scenarios. As a natural result, the test system combines *inefficient* test scenarios using cooperative learning module. This results in a combined test scenario that is time consuming and lacks the power to reveal more errors. Hence, the high cost function of the test system compared to the corresponding human participants.

To conclude, the delivered results do match with empirical observations, done in the last 25 years, which are summarized in Itkonen et al. (2009). Furthermore, the obtained results fit to the recent study in Kasurinen et al. (2009), in which they concluded that:

> "Testing should be executed by specialized personnel."

Though the main benefits that lie behind the usage of manual testing, it admittedly suffers from putting less emphasis on the documentation. Furthermore, classical test systems lack the power to trace-back what has been manually tested that could lead to overlooking significant test domains.

Considering the developed test system, unlikely to classical ones, it possesses the ability to systematically trace what is manually tested due to the online-observation module. This way, the achieved test coverage can be easily identified and, accordingly, the probability to overlook test domains is significantly decreased. Besides, one more non-trivial contribution is the capability to diagnose non-reproducible errors to identify error trigger(s).

Moreover, similar contributions like Gras et al. (2006) acquire experts' knowledge through an oral interview, which formulates an expensive burden. In addition, the fuzzy theory by Zadeh (1973) states that intuitive decision, taken by humans does not strictly stick to the law of logic, consistent models, or mathematics. As a result, humans usually verbalize their experiences quite vague and imprecise that could, therefore, lead to inconsistent training sessions.

Referring to the proposed test system in this Chapter, human testers are observed during test sessions with neither restrictions on their testing art, nor having to verbalize their test strategies. This would, therefore, lead to natural and consistent training sessions. In this context, it is worthwhile to highlight the ability of the test system to aggregate diverse strategies to maximize individual benefits.

Adding to this, the planning module spells the role to bridge the gap between manual and automatic testing. According to a pre-defined rule-based optimization algorithm, the planning module executes, supervises, and optimizes self-generated test scenarios. Hence, the developed test system frees test practices from any cognitive biases. This idea fits also to the results, obtained by Berndt et al. (2003), in which the final conclusion states:

> Combining both human and machine generated test cases from multiple sources may be the most robust strategy for many testing initiatives.

One more significant contribution of the developed test systems is its ability to partially generate the specifications of the DUT. Then, according to pre-designed criteria, various techniques can be applied to generate test scenarios based on the generated specifications.

Hereby, the last significant contribution is the ability of the test system to transfer the knowledge, learned in one test situation, to be used in similar ones. In fact, the central contribution of the generalization module is to capture a portion of functionalities that are relevant to a certain component in the DUT.

Finally, unlike most contributions that adopt case studies for evaluation purposes, this contribution validated the proposed idea via real-world experiments, away from any simulations that do not necessarily spell the reality.

Though the above addressed significant contributions, the developed test system suffers from diverse limitations. Indeed, the proposed idea demands the employment of skilled personnel in the training sessions. Yet, up to the knowledge of the author, there is *no* one unique-consistent definition of the term "skilled personnel". Consequently, this makes the selection process of participants quite vague and opposed to human biases. Apart from that, skilled personnel in test domains are quite expensive.

Furthermore, the introduced test system suffers from its limited domain of applicability. Indeed, it is too expensive to be applied in unit testing. Rather, it is only applicable for distributed systems like infotainment systems, driver-assistant systems, distributed aerospace units, etc. Furthermore, the addressed approach loses a reasonable portion of its usability, if the changes from a software version to the follower one are huge, e.g. Graphical-User-Interfaces (GUI).

One additional limitation is the non-ability of the system's designer to formally reason the motivations behind test strategies, adopted by skilled personnel. This is practically valid since humans generally employ heuristics and rules-of-thumbs to take intuitive-based decisions.

And, obviously, it is *not* practically possible to get an insight into human brain during testing. Adding to this, cooperative learning module decreases the probability to bias the learning process, but it does *not* lead to a commitment that there is no bias in learning processes.

References

Bach, J.: A framework for good enough testing. IEEE Comput. Soc. **31**(10), 124–126 (1998). The proceedings of the IEEE computer society

Berndt, D., Fisher, J., Pinglikar, L.J., Watkins, A.: Breeding software test cases with genetic algorithms. In: The Proceedings of the 36th Annual Hawaii International Conference on System Sciences, Hawaii, USA, IEEE Computer Society (2003)

Bezerra, E.A., Vargas, F., Gough, M.P.: Improving reconfigurable systems reliability by combining periodical test and redundancy techniques: a case study. J. Electron. Test.: Theory Appl. **17**(2), 163–174 (2001)

Dalal, S.R., Mallows, C.L.: Some graphical aids for deciding when to stop testing software. IEEE J. Sel. Areas Commun. **8**(2), 169–175 (1992)

Denaro, G., Mariani, L., Pezze, M.: Self-test components for highly reconfigurable systems. Electron. Notes Theor. Comput. Sci. **82**(6), 89–98 (2003)

Eltaher, A., Form, T., Ayeb, M., Maurer, M.: A generic architecture for hybrid intelligent test systems. In: The Proceedings of the 7th IEEE International Conference on Cybernetic Intelligent Systems, September, pp. 9–10. London, UK (2008)

Eltaher, A., Maurer, M., Form, T., Ayeb, M.: Agents learn from human experts: an approach to test reconfigurable systems. In: The Proceedings of the IEEE International Conference on Systems, Man, and Cybernetics, October, pp. 11–14. Texas, US (2009)

Gemoets, L., Kreinovich, V., Melendez, H.: When to stop testing software? a fuzzy interval approach. In: The Proceedings of the 1st International Joint Conference of the North American Fuzzy Information Processing Society, IEEE Computer Society (1994)

Gras, J.J., Gupta, R., Minana, E.P.: Generating a test strategy with Bayesian networks and common sense. In: The Proceedings of the Testing: Academic and Industrial Conference—Practice and Research Techniques, IEEE Computer Society (2006)

Itkonen, J., Mäntylä, M.V., Lassenius, C.: How do testers do it? An exploratory study on manual testing practices. In: The proceedings of the 3rd International Symposium on Electrical Software Engineering and Measurement, IEEE Computer Society (2009)
Kaner, C.: The impossibility of complete testing. Law Softw. Qual. Column, Softw. QA Mag. **4** (1997) pp 1–16
Kasurinen, J., Taipale, O., Smolander, K.: Analysis of Problems in Testing Practice. J. Asia-Pacific Softw. Eng. Conf., IEEE Comput. Soc. (2009) pp 309–315
Levendel, Y.: Using untampered metrics to decide when to stop testing software. In: The Proceedings of the IEEE International Conference on EC3-Energy, Computer, Communication and Control Systems, vol. 2, pp. 352–356 (1990)
Ngamsaowaros, N., Sophatsathit, P.: A novel framework for test domain reduction using extended finite state machine. In: The Proceedings of the 2nd International Conference on Software Engineering Advances, France, IEEE Computer Society (2007)
Söffker, D.: From human-machine interaction modeling to new concepts constructing autonomous systems: a phenomenological engineering-oriented approach. J. Intell. Robot. Syst. **32**(2), 191–205 (2001)
Wang, Q., Wang, S., Ji, Y.: A test sequence optimization method for improving fault coverage. In: The Proceedings of the 2nd IEEE International Conference on Information Management and Engineering, pp. 80–84, Chengdu (2010)
Zadeh, L.A.: Outline of a new approach to the analysis of complex systems and decision processes. IEEE Trans. Syst. Man Cybernet. **3**(1), 28–44 (1973)

MIX
Papier aus verantwortungsvollen Quellen
Paper from responsible sources
FSC® C105338

If you have any concerns about our products,
you can contact us on
ProductSafety@springernature.com

In case Publisher is established outside the EU,
the EU authorized representative is:
Springer Nature Customer Service Center GmbH
Europaplatz 3, 69115 Heidelberg, Germany

Printed by Libri Plureos GmbH
in Hamburg, Germany